SAE EDGE™
RESEARCH REPORT

Unsettled Issues Concerning Integrated Vehicle Health Management Systems and Maintenance Credits

Ravi Rajamani, PhD
Principal Consultant, drR2 consulting

EDGE DEVELOPMENT TEAM

Bal Annigeri, *Pratt & Whitney (Retired)*

Mark Davis, *Sikorsky (Retired)*

Bill Heliker, *Federal Aviation Administration*

Ian Jennions, *Cranfield University*

Marcus Labay, *Federal Aviation Administration*

David Piotrowski, *Delta Air Lines*

Pierre-Charles Rolland, *Airbus*

Guilherme Torres, *Embraer*

Rhonda Walthall, *Collins Aerospace*

SAE INTERNATIONAL®

Warrendale, Pennsylvania, USA

About the Publisher

SAE International® is a global association of more than 128,000 engineers and related technical experts in the aerospace, automotive, and commercial-vehicle industries. Our core competencies are lifelong learning and voluntary consensus standards development. Visit sae.org

SAE EDGE™ Research Report Disclaimer

SAE EDGE™ Research Reports focus on topics that are dynamic, in which knowledge is incomplete, and which have yet to be standardized. They represent the collective wisdom of a group of experts and serve as a practical guide to the reader in understanding unsettled subject matter. They are not meant to provide a recommended practice or protocol. The experts engaged have contributed their own thoughts and points of view, and these are not the positions of the institutions or businesses with which they are affiliated. A professional writer has collectivized their input; there is no one contributor's perspective being advanced but rather that of a community of practitioners. SAE EDGE™ Research Reports are the property of SAE International, and SAE alone is responsible for their content.

About This Publication

SAE EDGE™ Research Reports provide state-of-the-art and state-of-the-industry examinations of the most significant topics in mobility engineering. SAE EDGE™ contributors are experts from research, academia, and industry who have come together to explore and define the most critical advancements, challenges, and future direction in areas such as vehicle automation, unmanned aircraft, Internet of Things and connectivity, cybersecurity, advanced propulsion, and advanced manufacturing.

Related Resources

SAE MOBILUS® Automated & Connected Knowledge Hub
https://saemobilus.sae.org/automated-connected/

SAE Team

Frank Menchaca, Chief Product Officer

Michael Thompson, Director, Standards, Information, and Research Publications

Monica Nogueira, Acquisitions Director

Beth Ellen Dibeler, Product Manager

William Kucinski, Managing Technical Editor

EPR2020006
ISSN 2640-3536
e-ISSN 2640-3544
ISBN 978-1-4686-0183-1

To purchase bulk quantities, please contact: SAE Customer Service

E-mail: CustomerService@sae.org
Phone: 877-606-7323 (*inside USA and Canada*)
724-776-4970 (*outside USA*)
Fax: 724-776-0790

https://www.sae.org/publications/edge-research-reports

About the Editor

Dr. Ravi Rajamani is an independent consultant in the aerospace and energy sectors. He has many years of experience in the application of systems engineering principles and data analytics and model-based methods to controls, diagnostics, and prognostics, especially for propulsion systems. He has authored or coauthered five books, including *Electric Flight Technology: The Unfolding of a New Future*. In addition, Dr. Rajamani is the author of many book chapters, journal articles, conference papers, and patents. Prior to his current job, Ravi worked at Meggitt, United Technologies Corporation, and the General Electric Company. He has a PhD from the University of Minnesota and an MBA from the University of Connecticut. His earlier degrees were BTech from the Indian Institute of Technology, Delhi, and an MSc from the Indian Institute of Science, Bangalore. He is active within various SAE technical committees dealing with prognostics and health management (PHM) and electric propulsion. He is also active in the PHM Society, serving on its board of directors. He has been elected as Fellow of SAE International and of the Institution of Mechanical Engineers. He currently serves as the Editor-in-Chief of the SAE International Journal of Aerospace. In addition, he has a research appointment at the University of Connecticut and is a visiting Professor at Cranfield University.

contents

Unsettled Issues Concerning Integrated Vehicle Health Management Systems and Maintenance Credits

Abstract

The "holy grail" for prognostics and health management (PHM) professionals in the aviation sector is to have integrated vehicle health management (IVHM) systems incorporated into standard aircraft maintenance policies. Such a change from current aerospace industry practices would lend credibility to this field by validating its claims of reducing repair and maintenance costs and, hence, the overall cost of ownership of the asset. Ultimately, more widespread use of advanced PHM techniques will have a positive impact on safety and, for some cases, might even allow aircraft designers to reduce the weight of components because the uncertainty associated with estimating their predicted useful life can be reduced. The journey to that end will not be easy, but we believe that it will be successful. The entire idea of diagnostics and prognostics is to look at sensed data along with the knowledge about system operations from accurate models, and develop an estimate of whether the system is behaving in an acceptable manner. Even when operations are within a range of acceptable behavior, IVHM systems can predict future behavior of the systems from estimated deviations and trends. For example, vibration levels in an engine might be acceptable within engineering limits of operation but might start trending consistently in the "wrong" direction. Using this information to predict the condition of the engine in the future would constitute a prognostic estimate that can be used to schedule appropriate upcoming maintenance actions. In general, maintenance practices in the aviation world depend on regular inspections to ensure that systems are functioning correctly. These maintenance practices are enhanced by condition indicators that can trigger alarm conditions that require immediate or imminent action. Until quite recently, the use of IVHM systems to either increase inspection intervals or eliminate them in favor of fully condition-based maintenance (CBM) has not been deemed acceptable by regulatory authorities. In this report we will outline a few developments in recent times that point to a more favorable future for IVHM systems in affecting established maintenance

RAVI RAJAMANI, PHD
Principal Consultant, drR2 consulting

Edge Development Team

Bal Annigeri, *Pratt & Whitney (Retired)*
Mark Davis, *Sikorsky (Retired)*
Bill Heliker, *Federal Aviation Administration*
Ian Jennions, *Cranfield University*
Marcus Labay, *Federal Aviation Administration*
David Piotrowski, *Delta Air Lines*
Pierre-Charles Rolland, *Airbus*
Guilherme Torres, *Embraer*
Rhonda Walthall, *Collins Aerospace*

ISSN 2640-3536

practices. We will discuss how standard maintenance procedures are developed, who the various stakeholders are, and - based on this understanding - outline how new PHM systems can gain the required approval to be included in these standard practices. There have been a few limited successes in this field already, and we will discuss the lessons learned in developing these systems. Finally, we will review the progress that the structural health management (SHM) community has made, and continues to make, to change the way the industry regards automated SHM systems.

Before we go further, a quick note on the terminology. I define PHM as the engineering discipline that teaches the techniques and the science behind the development of diagnostics and prognostic systems. The most comprehensive (highest level) embodiment of PHM functionality when applied to a vehicle is an IVHM system. Below this are more specialized systems such as Engine/Equipment Health Management (EHM) System, SHM, etc. However, because of the somewhat fluid nature of this discipline, the abbreviations PHM or IVHM are used interchangeably, and that would be the case in this report as well.

Today, the basis for all maintenance practices in the commercial aviation sector is a close partnership between the industry and regulatory authorities. This has been codified by a rigorous set of practices developed in the 1960s to help with the maintenance of complex modern aircraft, such as the Boeing 747. The advantage of this system is that it establishes a well-understood and well-accepted path to developing maintenance procedures that all stakeholders agree with. The disadvantage is that novel ideas - such as PHM - take longer to be accepted and must overcome a lot of skepticism at the beginning. In a safety-conscious industry such as ours, this is exactly the way it should be, but we need to reconcile with the fact that the time it will take for the industry to accept novel ideas will be longer than usual. One key to doing this successfully will be to align all new initiatives with the maintenance review board (MRB) process. The MRB is constituted for each aircraft during system certification, and approves the maintenance plan which is then used by the operator to develop detailed maintenance procedures. We will discuss the details of this process and recommended steps that will need to be taken to include IVHM into these procedures. Because aircraft operate around the world, there is a need to coordinate the work of these MRBs in different countries. This is done via the International Maintenance Review Board Policy Board (IMRBPB). We will discuss how this organization can be used to help change the procedures regarding the use of an IVHM system for maintenance credits.

Maintenance credits are the key to making diagnostics and prognostics systems effective. If, by the use of an IVHM system, the maintainer is able to earn maintenance credits, it means that the maintenance procedures can be modified in such a way as to save costs. Inspection intervals can be lengthened or eliminated, and on-condition procedures can be replaced by condition-based procedures. In terms of fault isolation, advanced diagnostics and prognostics ideally target a line-replaceable unit (LRU) for the isolation. This allows the LRU to be swapped out quickly so that the mission can proceed without a significant delay. The LRU can then be returned to the shop for further analysis and repair. But enhanced IVHM systems can go further and isolate faults to specific components. Knowing exactly where the fault is can help reduce turnaround times in repair shops. This is an important benefit of the use of such systems, but we will not delve into this here because it does not pertain to maintenance credits. It is important to note that the authorities will only approve changes to established maintenance procedures if they can be convinced that these changes will not compromise continued airworthiness in any way.

The rotorcraft industry has been particularly successful in allowing operators to earn maintenance credits by using a limited number of health monitoring systems in the operation of some advanced helicopters. This has allowed the operators to reduce their aftermarket costs. We will discuss this experience and list some of the lessons learned that can be transferred to the commercial aviation sector. Health and Usage Monitoring System (HUMS) were first introduced (in fleet-wide deployment) in helicopter fleets servicing the North Sea oil rigs. This allowed the helicopter operators to catch problems with the transmission system before they led to nonrecoverable failures. HUMS technology was then adopted by the defense industry and civilian fleets in other parts of the world, notably in the USA. The rotorcraft part of the Federal Aviation Administration (FAA) successfully published a miscellaneous guidance document as an Appendix to an advisory circular (AC 29-2C), termed MG15. This guidance, called "Airworthiness Approval of Rotorcraft Health Usage Monitoring Systems," laid out the steps that an original equipment manufacturer (OEM) could take to obtain credits using HUMS. MG15 allowed some OEMs to institute procedures whereby operators could get relief on part retirements based on recorded usage. The fixed-wing community is in the process of doing something similar. We will look at a proposed advisory circular (AC 43-218), released for public comments, that gives guidance for these practices. The responses from the industry on the draft version of this document give a sense of the unsettled issues regarding the use of aircraft health management systems in aircraft maintenance. In a number of ways, the SHM community seems to be taking a leading role in developing procedures for monitoring aircraft structures while keeping the regulatory authorities closely in the loop. We will devote some time to discussing their experience which can give us some pointers on how other stakeholders can advance their own cause.

It is hoped that, at the end of the day - with the discussion of these unsettled issues - we will be closer to seeing a clearer path for IVHM systems to be installed routinely on commercial aircraft and for the benefits to be enjoyed by operators and suppliers alike.

Introduction

State of the Industry

This is a good time to be a PHM professional in the aviation industry. Companies are realizing that IVHM systems can deliver benefits, at times with minimal investments in development. There are definitely certification challenges associated with the introduction of new PHM capabilities, and we will tackle those in this report. Additionally, these systems do not have to fundamentally change the way aircraft are designed and operated. With some additional sensors and analysis, IVHM systems can deliver more lifecycle savings compared to other systems that might require a fundamental redesign of the aircraft. The advent of advanced digital technologies in the form of artificial intelligence (AI) and machine learning (ML) algorithms have thrown open the door for developers to create an array of advanced data-analysis functions that have improved the performance of these IVHM systems considerably. Some of the drawbacks of these systems, such as unacceptable false-alarm rates and missed-detection rates have decreased, and their reliability has increased. Yet we have not been very successful in using these systems to change the way maintenance is carried out (i.e., actually updating maintenance procedures). It is true that a lot of progress has been made, but much of that has been in the military world where, in general, certification requirements are not that onerous.

This is not to say that CBM practices have not been introduced into the world of aviation maintenance, repair, and overhaul (MRO). It is just that these have been few and far between. In a world where the standard practice still consists of conducting inspections at regularly spaced intervals, it is difficult for the maintainers to trust modern monitoring technologies. Most of the repairs are carried out based on the results of these inspections. Some condition indicators, such as for engine vibrations or oil-debris monitoring on some aircraft, do allow these intervals to be lengthened (i.e., the IVHM systems allow the operator to "earn" maintenance credits). But these are the exception and not the rule. Some of these systems will be described in this report so that we can learn from their principles, and see what can be applied more universally.

Having said that, the picture is not all that discouraging. One of the most important developments in this area is unfolding right now, because the FAA has issued a draft AC which addresses the topic of IVHM systems. It seeks to establish guidance for how applicants can approach the FAA to obtain maintenance credits using IVHM systems [7]. The draft AC 43-218, published in August 2019, has been reviewed widely by the aviation industry, and as this SAE EDGE™ Research Report is being finalized, the FAA has responded to all comments and is ready to publish the AC. The discussion here is based on a review of the pre-publication draft version, which was shared with this author. It is encouraging to see the regulators take this first step, because without the active support of the authorities, substantive changes in the maintenance practices will be impossible - especially when it comes to advanced practices such as PHM. It should be noted that in this area, the rotorcraft industry is ahead of its commercial transport aircraft counterpart. Back in 2003, the rotorcraft authorities added a miscellaneous guidance appendix (MG15) to their certification document AC 29-2C (Certification of Transport Category Rotorcraft [6]), where they laid out guidelines for how rotorcraft manufacturers could obtain credits for the use of HUMS. We will discuss what the rotorcraft industry has done with this topic, and suggest some steps the commercial fixed-wing industry can take to get on par.

Aircraft maintenance has a rich history, with many diverse practices existing for MRO across different operators. There were few standards to help guide operators on how to develop a maintenance program. This changed in the 1960s, when the airlines and OEMs came together to develop standards for maintaining aircraft. The classic methods of inspections and replacements based on fixed intervals gave way to procedures based on more scientific findings on how the reliability of different components (i.e., how they fail and how their condition deteriorates over time).

While not comprehensive, Figure 1 gives some idea of the history of how maintenance practices have evolved in the aerospace world. One of the key developments was the scientific study of failure mechanisms for various aircraft systems and the optimization of maintenance actions based on this knowledge. This was the basis for reliability-centered maintenance (RCM) that was pioneered by the US Department of Defense (DoD), as well as large operators such as United Airlines. Nowlan and Heap, from United Airlines, authored a book in 1978 that laid out the principles of RCM [18]. Implementing these principles resulted in substantial benefits for the industry. For example, according to Nowlan and Heap, United Airlines estimated that they spent 66,000 maintenance man-hours (MMH) for a major structural inspection of the Boeing 747 using the new maintenance practices to get to a certain major maintenance interval. In contrast, they used to spend over four million MMH to perform the same inspections for the smaller and less complex Douglas DC-8. This huge reduction was accomplished without any compromise to the operating reliability of the aircraft. A clearer understanding of the failure mechanisms allowed the operator to design inspection intervals appropriate to the reliability of the components, rather than scheduling them based on historical practices. Because the committee that was constituted to author these guidelines was called Maintenance Steering Group (MSG), that became the name of the documents as well. Even though the name of the group changed to Maintenance Programs Industry Group (MPIG) in 2008, the most recent edition of the document is still called MSG-3. The first version of these guidelines was published in 1968, primarily to benefit Boeing 747 maintainers; and the latest version, published in 1980, incorporates learning from the implementation of RCM principles.

While this development really helped revolutionize maintenance practices in the airline world, the larger goal of CBM was still quite distant. The basic philosophy in aircraft

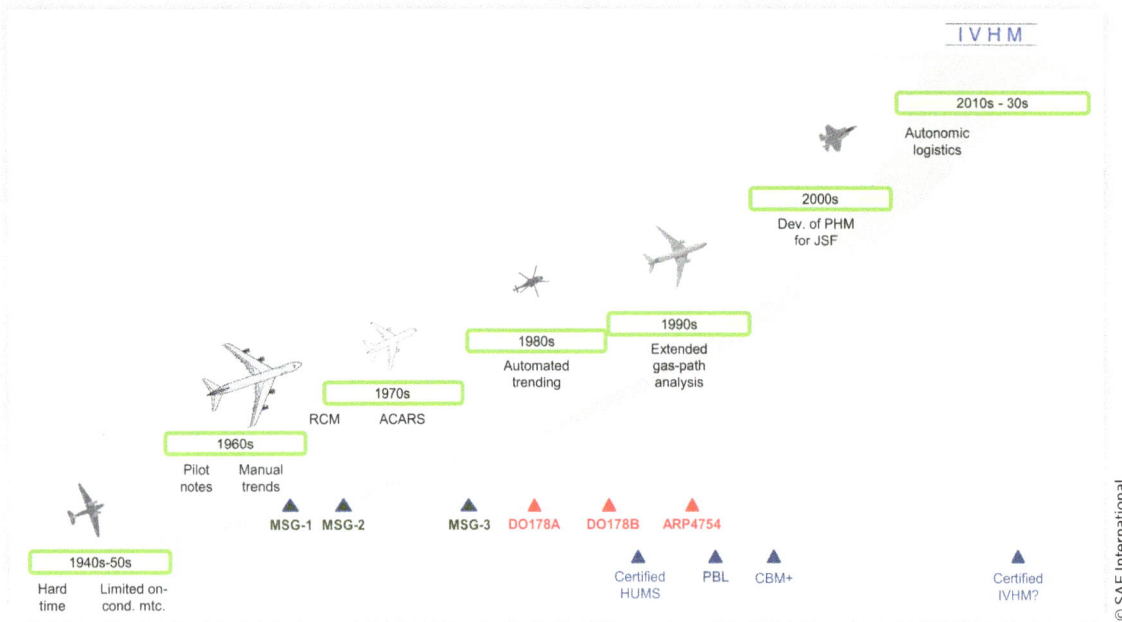

maintenance was - and still is - centered on performing regular maintenance inspections and checks. Because of the highly safety-critical nature of the aviation industry, the chances of it moving away wholesale from this philosophy is low. But, to be fair, that is the right attitude for such a risk-averse industry to have. The move to a more condition-based future for maintenance will have to be evolutionary rather than revolutionary.

It should be clear that those advocating an increase in the use of PHM functionality are perfectly happy with a slow growth in the use of this technology. Their idea would be to look at specific maintenance items and figure out whether diagnostic and prognostic functions can be employed that allow them to be maintained in a more "condition-based" manner rather than based solely on their operating time. Any change would then be introduced slowly, and in parallel with existing practices. Finally, when enough evidence is gathered to justify a more permanent switch, the move would be almost inevitable because of the accrued benefits, which, by then, would be plainly evident.

In this report, we will lay out some of the issues that need to be resolved before this can become a reality. The first order of business is to understand how the aviation world is set up today to plan and execute on maintenance practices. This will go into some of the new developments within the regulatory world to see how they affect the path forward. Next, we will look at the rotorcraft industry and consider how they have managed to get further ahead with the applications of PHM principles to obtain maintenance credits. Even with rotor-craft, these practices are still evolving, and we will discuss some of the open issues. Finally, we will look at some concrete examples of PHM functions that have been implemented in

the industry and which, with some judicious application of the recommendations in this report, can be taken further.

Before going further, however, a note on Development Assurance Levels (DAL) or system criticality is required. Failure conditions in an (airborne) aircraft system are classified according to their consequences. As defined in FAA document AC 25.1309-1A, there are four consequences of a failure: Catastrophic, Hazardous, Major, and Minor, with No Safety Effect added later [9]. To mitigate each type of failure, a certain level of development rigor is needed, and that is what the DAL categorization is all about. If the failure has a catastrophic effect, then the level of rigor needed to mitigate it would be of the highest level (i.e., Level A), and we can go down successively to lower levels and end at Level E for a failure that has no effect. In the latest version of ARP4754 (i.e., ARP4754A [27]), a distinction has been made between functional DAL (FDAL) and item DAL (IDAL). The latter refers to software and electronic hardware items associated with airborne systems, and the former to system functions at the highest level. The definitions of the failure consequences are given in Figure 2, with the rightmost column listing the probabilities associated with each failure in IDALs. It should be noted that these numbers refer to failures per flight hour (FH). These numbers can be derived from the reliability numbers associated with the components and systems, as well as from similar systems on older aircraft.

The reason for standardizing these definitions is to ensure that safety analysis can be implemented in a common manner across all programs and to be able to standardize the design process. This way of characterizing the assurance levels has been most useful in establishing how software and electronic hardware systems are designed. While ARP4754A

FIGURE 2. **Criticality levels.**

Definition of failure condition	Description	Probability of occurrence (per FH)
Catastrophic	Failure conditions which would prevent continued safe flight and landing	$P < 10^{-9}$
Hazardous/ Severe-Major	Failure conditions which would reduce the capability of the aircraft or the ability of the crew to cope with adverse operating conditions	$P < 10^{-6}$
Major	Failure conditions which would reduce the capability of the aircraft or the ability of the crew to cope with adverse operating conditions to the extent that there would be a significant reduction in safety	$P < 10^{-5}$
Minor	Failure conditions which would not significantly reduce aircraft safety, and which would involve crew actions that are well within their capabilities	$P < 10^{-3}$
No effect	Failure conditions which will have NO effect	None

© SAE International.

discusses the system-level process for the design of an aircraft, the specific guidance for the design of software and electronic hardware is presented in DO178 (latest version C [20]) and DO254 [21], respectively. These documents have been developed by the RTCA in the US and EUROCAE in Europe. In DO178, if a certain failure of a system was deemed to be catastrophic, then the software that mitigated that failure would have to be designed at an IDAL of A. This put a whole lot of constraints on the team developing the software, such as independence of the design and the review processes, modified condition/decision coverage, configuration management, etc., that made the software development process very expensive. As outlined in Figure 2, the DAL could be successively lowered as the failure become less and less consequential. For more details the reader is directed to ARP4754A and DO178C.

Unsettled Issues Concerning IVHM Systems and Maintenance Credits

There are several unsettled issues with the use of the IVHM system in the maintenance world, and we will consider some in this report. The focus of this report is on how IVHM systems can be used to obtain maintenance credits. A maintenance credit is earned when the maintainer gets to change the maintenance procedure of any part or system in an aircraft in a way that leads to a cost savings without compromising the continued airworthiness of the aircraft. Typically, such credits would be obtained by lengthening the inspection interval or eliminating an inspection entirely, or by allowing a part to continue to operate on the aircraft beyond what it would normally have been allowed to. All these procedures would mean, by definition, that the new maintenance process has deviated from the established instructions for continued airworthiness (ICA). If this change is initiated at the beginning

of the life of the aircraft, i.e., the PHM function is built into the initial design during type certification (TC), then the changes will have to be approved at the very beginning and incorporated into the maintenance plan. If, on the other hand, the PHM function is retrofitted after the aircraft is operational, then the situation becomes more complicated. Some modification to the approved maintenance plan would have to be developed, which could result in a supplemental type certificate (STC) that would need to be issued for any new hardware.

The report will consider these three main topics:

1. Regulations related to application of PHM systems
2. The rotorcraft experience
3. Specific PHM use cases in commercial aviation

Regulations Related to Application of IVHM Systems

When compared to any other transportation sector, or even most other industries, the commercial aviation sector has established safety standards which are at much higher levels. The strikingly small number of accidents and incidents involving airliners is adequate proof that such a cautious approach was the correct one. Hence, in commercial aviation, it is a given that regulators have a fundamental responsibility of ensuring that anything the applicants propose in terms of maintenance credits is thoroughly vetted and validated before approval. So it is important to understand the process by which any novel maintenance procedure could be introduced into standard practice.

The main unsettled issue regarding this topic is the readiness of authorities to accept automated systems that monitor system health to direct maintenance actions. To understand this better, let us look at how maintenance procedures are developed in the first place. In describing this, we will concentrate on commercial aviation. While the engineering principles that govern this process are identical for commercial and military aircraft, the regulations governing the approval of maintenance procedures for the former are stricter and more prescriptive. The content in this section has been informed by discussions with Bill Heliker and Marcus Labay from the FAA, David Piotrowski from Delta, and Guilherme Torres from Embraer.

Beginning in the 1960s, the commercial aviation industry realized that it would be in the collective interest of all stakeholders in this area (airlines, OEMs, suppliers, maintainers, and the regulators) to codify the principles of scheduled maintenance in a more formal manner. The Air Transport Association (ATA, which later became Airlines for America, A4A) constituted the MSG to develop procedures that would define how a large aircraft would be maintained. The first target of this exercise was the Boeing 747. As mentioned above, this exercise of regularizing the scheduled maintenance practices resulted in a significant reduction in the maintenance burden for operators - by more than 95 percent in some cases. The MSG process, which has now gone through two major iterations, and countless minor ones, is a living document. It provides the "tools" for the industry to construct the maintenance plan. Its goal is to lay out, now using the principles of RCM, what the scheduled inspection intervals are for different maintenance significant items (MSI) and structural significant items (SSI) to ensure that there are no incipient failures. These are obviously not the only causes for maintenance in aircraft. Bird strikes, sensor malfunctions, and other maladies disrupt operations. These are not covered by the scheduled maintenance practices; procedures for handling these are developed separately.

We are focusing on the standard process, because if any IVHM system is adopted by the industry, it will have to go through this process to be acceptable for general use. Because of strict safety requirements, the entire process has evolved to have many checks and balances. While some systems still fall through, for the most part this maintenance program development process has served the industry very well over the last six decades.

Figure 3, taken from AC 121-22C [8], illustrates the standard process for developing a maintenance plan for a new aircraft. It starts at the bottom left with the TC of an aircraft.

FIGURE 3.   **MRB process flowchart.**

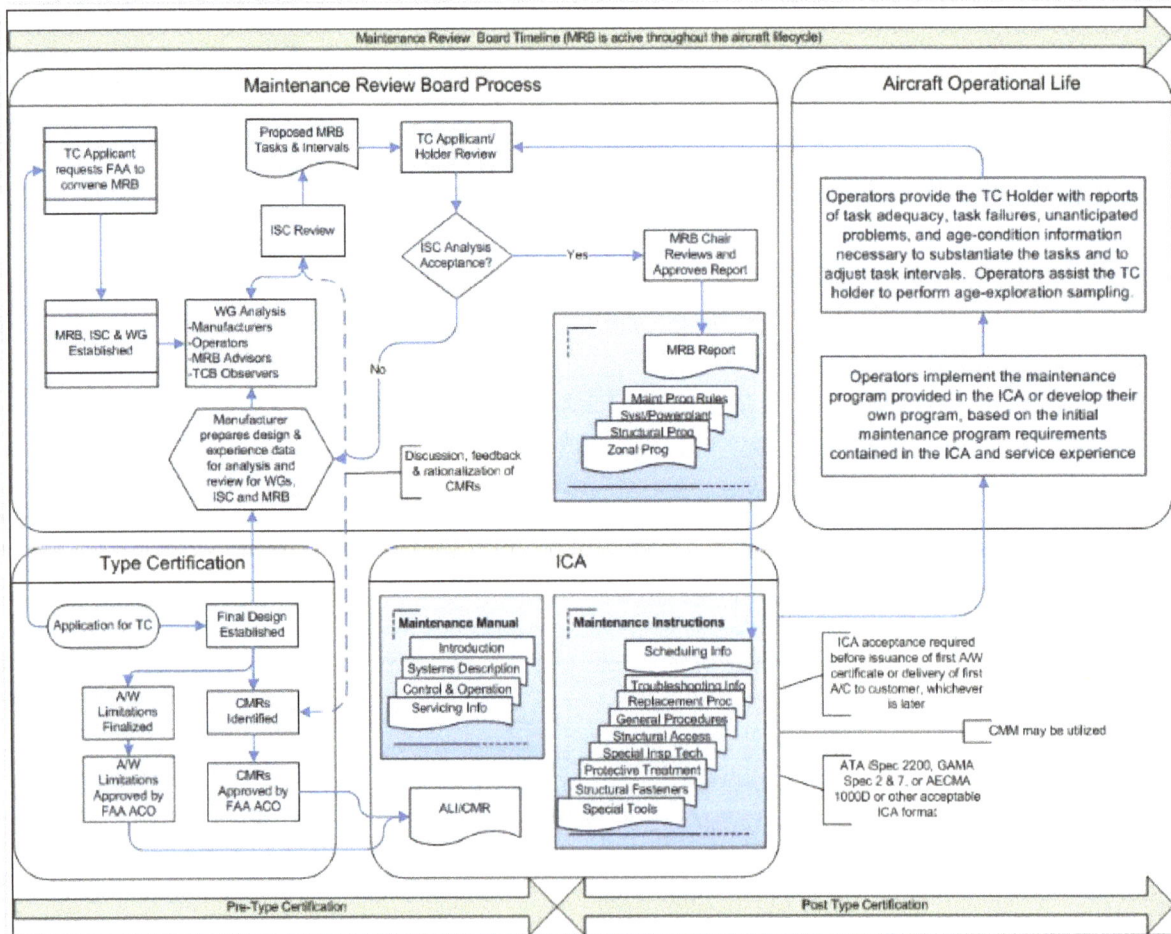

Reprinted from Ref. [8].

The maintenance planning process does not have to wait for the TC to be issued; the applicant can request for the MRB to be convened when the main critical tasks related to maintenance (i.e., the Certification Maintenance Requirements) have been completed during the TC process. The MRB is chaired by a representative from the airworthiness authority, which, in the US, is FAA Flight Standards or, in other countries, their National Airworthiness Authority (NAA). This committee consists of representatives from the applicant company and any major suppliers (e.g., the engine OEMs) and the aircraft operators involved with the specific aircraft. The result of all these deliberations is one of the most important documents ensuring the airworthiness of an aircraft: the MRB Report (MRBR). This represents the approved high-level plan for maintaining the aircraft and satisfies the prerequisites for developing the detailed ICA as required by 14CFR 21.50(b), which states

(b) The holder of a design approval, including either a type certificate or supplemental type certificate for an aircraft, aircraft engine, or propeller for which application was made after January 28, 1981, must furnish at least one set of complete Instructions for Continued Airworthiness to the owner of each type aircraft, aircraft engine, or propeller upon its delivery, or upon issuance of the first standard airworthiness certificate for the affected aircraft, whichever occurs later. The Instructions for Continued Airworthiness must be prepared in accordance with §§23.1529, 25.1529, 25.1729, 27.1529, 29.1529, 31.82, 33.4, 35.4, or part 26 of this subchapter, or as specified in the applicable airworthiness criteria for special classes of aircraft defined in §21.17(b), as applicable. If the holder of a design approval chooses to designate parts as commercial, it must include in the Instructions for Continued Airworthiness a list of commercial parts submitted in accordance with the provisions of paragraph (c) of this section. Thereafter, the holder of a design approval must make those instructions available to any other person required by this chapter to comply with any of the terms of those instructions. In addition, changes to the Instructions for Continued Airworthiness shall be made available to any person required by this chapter to comply with any of those instructions.

Given an approved MRBR, a detailed maintenance plan can be developed, which results in maintenance manuals and instructions that the OEM puts together with help from key suppliers, such as the engine OEM. This is the Maintenance Planning Document (MPD) which is not "approved" by the authorities, but "accepted." This will contain not only scheduled MSIs and SSIs but also other maintenance tasks (e.g., those required during the warranty period). Once the MPD is handed over to the operator, detailed maintenance procedures can be developed. This is very specific to each operator. It is known as the maintenance manual and instructions, or the Approved Maintenance Program (AMP), and it governs the day-to-day operations of the aircraft maintainer. The MPD is typically four times the size of the MRBR, and the

detailed maintenance instructions manual is even larger; of course, the sizes of these documents will vary based on the size and complexity of the aircraft. A regional jet probably will not have as many maintenance items to consider as a twin-aisle jumbo jet.

A simplified version of this detailed process is shown in Figure 4, which has been modified from a diagram presented by the European Aviation Safety Agency (EASA).

Now let us walk through the process of introducing a new maintenance procedure based on new experience, knowledge, or technology. This can be initiated by an OEM or the operator. An example of a significant change (one that affects an MSI or SSI - both of which are covered by MRB processes) introduced by an OEM would be brake monitoring. It is currently an MSI that requires a visual check of the brake pins to ensure that each brake stack has adequate remaining thickness. Let us assume that the brake supplier has come up with an automated process that allows this estimate to be made without the use of any inspection. This means that there is an automated system that assesses the health of the brake disks and reports their status to the aircraft. In addition, assume that the OEM has developed analytical tools that will ensure that a notification for replacing the brakes is automatically generated and all relevant parties notified. This innovation could lead to a complete elimination of visual inspections. While the visual check is not difficult, it can be time consuming and inconvenient, especially in inclement weather and low visibility conditions. A future scenario where the visual check has been eliminated and replaced entirely by an automated IVHM system is not unthinkable. Many companies have been working on such IVHM systems for a while now. Even mid-market automobiles have sensors that indicate when it is time to replace the brake pads. With the advent of electrical braking systems (e.g., on the Boeing 787 or the Airbus/Bombardier A220), this becomes even easier because the information about brake travel is readily available. Even this author is the co-inventor of a system that uses model-based techniques to estimate brake wear [4]. Such an automated brake monitoring system has been discussed in the industry for quite a while. Issue Paper 180 on the integration of aircraft health monitoring (AHM) systems into MSG-3 has outlined this as one of the examples of an AHM system that can replace visual inspections [14].

If such a brake monitoring system is developed during the design and TC stage, the entire process for obtaining these kinds of credits is straightforward. Brake monitoring would be included in the original application to the MRB. All relevant evidence needed to back up the claims would already have been generated during the design phase and included while getting the TC. The brake supplier and the OEM would have enough data to approach their MRB to put forth the case for a significant change to the maintenance procedure from past practices. If all the experts in the room agree, it would result in a new maintenance procedure where the maintenance credit would be baked into the MPD and the maintenance plan.

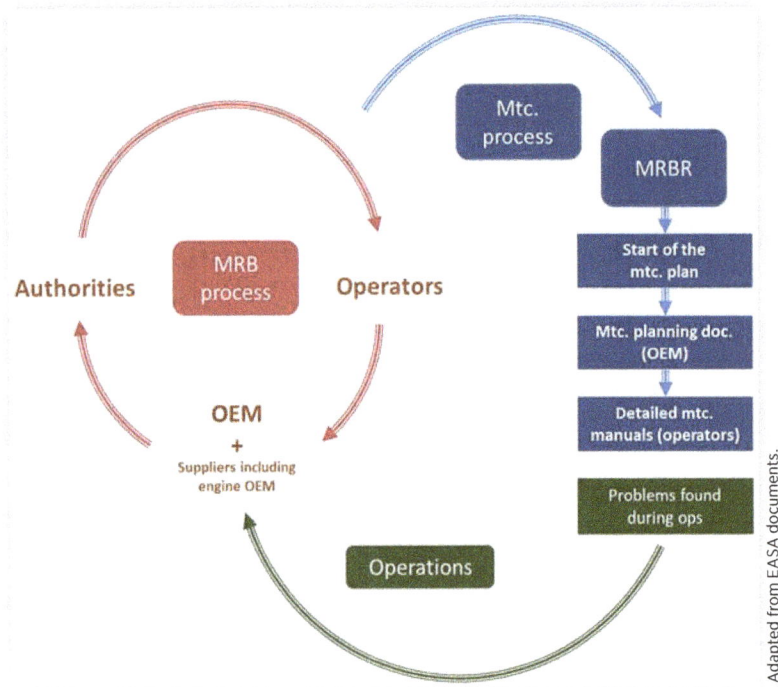

Adapted from EASA documents.

FIGURE 4. **The MRB process.**

The more interesting case is when this feature is a retrofit. For many IVHM systems, this would probably be the case because these systems are developed often as a result of cost-benefit analyses to address maintenance expenditures. The biggest issue in this case would be to prove to the authorities that any change to the maintenance practice would not in any way compromise the safety of the braking function. Because aircraft brakes are the primary means of stopping an aircraft (and not thrust reversers, which are optional), any novel feature or procedure affecting brake maintenance will be thoroughly scrutinized for potential issues.

Let us further assume that the originator of this request is (as in the previous case) the brake manufacturer. Then the process would not be too different than the one outlined previously. The MRB would be approached and a case would be presented to make a change to the maintenance procedure. Any changes to the aircraft itself would have to go through an STC exercise; whereas if the changes only involve some analysis, then an STC would not be necessary. For example, the system envisioned in the patent application cited above does not involve any new hardware, just offline models that would estimate brake wear based on aircraft usage.

The more interesting case is when the organization that wants the retrofit solution to be implemented is the *operator*. The operator would be the direct beneficiary of any credits obtained through the implementation of a new procedure, so it is likely that such changes would be initiated by the operator. In this case the route by which the function is approved will be more complicated. The first organization the (US) airline would contact would be the Certification Management Unit (CMU), which is located at the operator location. Because the change that is sought is beyond the

scope of the CMU, the FAA Airline Certification Office (ACO), which has oversight responsibility for the airline, will have to get involved. This is after the airline's design and engineering departments have done all the necessary engineering studies and validation tests, and have all the data that proves the efficacy of the new PHM function. Armed with this data, the ACO would be informed about the desire to change the maintenance procedure. Because this change in the brake monitoring procedure is significant, the airframer and the brake manufacturer would likely get involved. The goal, as in the previous scenario, would be to affect a change to the MPD.

Because the significance of the change would affect the MPD, it would have to flow through a change to the MRBR, which means that the operator will ask the MRB to be convened. Before the airframer or any of the suppliers are consulted, the MRB would determine the significance of the requested change and the magnitude of the maintenance action. This could lead directly to a change in the MRBR, or it could lead to the airframer (and any cognizant supplier) being called in, so that all aspects of the change could be considered. Once this is taken care of, the MRBR and the MPD would be altered to reflect the change in the scheduled maintenance practice. Visual inspections would be eliminated and the automated system, recommended by the operator, would replace it. This would lead to the operator changing its maintenance plan and its detailed procedures so that the projected benefits could be realized.

But what about the changes that have been made to the MPD? Will these be automatically flowed down to other operators who will be competing against the operator that developed the new procedure? This brings up the issue of the investment in research and development and in securing intellectual property leading up to this change. Given the nature of the aerospace industry, this might not be insignificant! Without a doubt, the operator would have invested a good deal of time and money to come up with this new feature. Which means that, if the MPD is modified so that other operators could benefit from this feature, there is a question about how the inventor will be compensated. It is not clear that anybody knows how these issues will be handled. This is an unsettled issue that needs to be resolved.

Regardless of where and how the change to the maintenance procedure originates, it is important to note that it will be an *alternate* means of compliance (AMOC). This means the original procedures will remain in place and the new procedure can be used as *another* method of achieving the continued airworthiness goals, if the operator so desires. The goal in all this is to ensure that the aircraft continues to maintain airworthiness regardless of how the maintenance procedures are changed.

As noted above, data will become a significant issue for all stakeholders in the future. In the aerospace world, as in all spheres of our existence, data is becoming the governing entity - the currency of the digital realm. The example of the big five tech giants has established that the organization that owns the data becomes king. There is a struggle being played out in the industry today where every entity involved in the data chain is staking out its claim to be the owner. The airlines have the clearest claim to this, since they fly the planes and actually own the asset and - barring any explicit service contracts - it is their data to use as they please. However, there could be two hurdles to this: data encryption and analytical knowledge. The OEM-supplied data system can be readily encrypted so that without the "decoder," the data cannot be read. Secondly, even if the data can be read, without the models or the analytical methodology (the "secret sauce" if you will), which only the OEMs can provide, the data by itself is useless. In the specific example outlined above, the airline may have been able to develop the algorithms needed to track brake wear, but often the physics is so complicated that nobody, but the designer, will be able to interpret the data correctly. The example of life-limited parts (LLP) that we present later in this report proves this point.

This is one reason why airframers and major suppliers alike are offering long-term service agreements to their customers. These contracts, which were popularized by Rolls Royce at one time as "power-by-the-hour" contracts, ensure that the OEMs - at least for the most comprehensive service offering - take on the responsibility for all maintenance on the product in return for a payment from the airline that is based on the operational hours. In return for this annuity, the responsibility for ensuring a guaranteed uptime on the product then falls on the shoulders of the OEMs. They maintain the products, and because data is essential for this to work, the operators are obliged to share it with the OEMS to help them with maintenance. Without such contracts, it is not clear who owns the data. This is a highly contentious issue in the aviation industry today, and it is one of the many unsettled issues that will need to be addressed before the industry achieves completely automated maintenance scheduling. To address the issue of digital data, SAE International has established a number of technical committees to develop consensus documents to support the industry as it transforms. The idea is to create procedures that have the support of all stakeholders so that the industry can move forward in a collaborative fashion. Other consortia and organizations such as A4A, International Air Transport Association, and International Civil Aviation Organization are also working to develop future digital data solutions for the commercial aviation industry. Much has already been done, but much more needs to be accomplished.

Our hypothetical aircraft brake example is probably something that is technically feasible with today's technology and could be implemented relatively quickly. Other examples that have been implemented or are being developed, include vibrations monitoring, oil debris monitoring (ODM), water wash prediction, usage-based lifing (UBL), and extended operations. But the grand goal of a complete CBM plan is far from becoming a reality. Gaining approval for any advanced PHM function is not easy. Leaving safety entirely in the hands of automated processes is - intrinsically - a scary prospect for an industry which is used to placing safety above all other considerations, even if all the evidence points to the new system being as safe as or safer than the extant process.

For example, the National Safety Council estimates that just over 40,000 people died in vehicle accidents in 2017 [17]. A total of four deaths have occurred over the last three years that involve self-driving vehicles. While 40,000 deaths per year elicits little in the way of public response, one death per year due to an automated system creates headlines. This is understandable to an extent, because somehow if there is a human in the loop, the death does not seem as arbitrary as when it occurs due to the malfunction of an algorithm.

A similar thought process might be at play when it comes to regulations regarding automated IVHM systems, where decisions are being taken by an algorithm based on data and without a human intervening to either validate or reverse this decision. Regulators tend to be more lenient in their outlook when there is a human in the decision-making loop, even though it is not clear that those humans can react any better than a machine in high-pressure scenarios.

Regardless of this, the process of obtaining relief from the authorities for monitoring instruments is a systematic process. Once the approval of the MRB is obtained, it will be reflected in the MRBR and the MPD, which will allow operators to incorporate the new maintenance procedure into their MRO plans.

A better process for the industry might be for the change to be recognized as a standard practice. This would mean that this change be discussed and approved at a level higher than the MRB. In 2006, the IMRBPB developed a process to track how such changes can be made, reviewed, and implemented. The IMRBPB is a regulatory body consisting currently of representatives from individual national MRBs from ten nations (the USA, Europe, Brazil, Singapore, China, Canada, Hong Kong, Japan, Australia, and the United Arab Emirates). The IMRBPB also meets annually with the MPIG, and the two work together, among other things, to update the MSG maintenance guidance. The mechanism for introducing new policies and procedures via the IMRBPB generally starts with the writing of an issue paper (IP). Any member of the IMRBPB or MPIG can introduce a candidate IP (CIP) to the IMRBPB. This could be a convenient way to approach the authorities to take cognizance of the benefits of the use of PHM in maintenance practices. The automated brake-monitoring system as a replacement for brake pin inspections could very well form the basis for such a CIP. These IPs are a means by which the IMRBPB communicates issues regarding aircraft maintenance to the larger aviation community. In fact, there are a couple of IPs that are already published or are being worked on which are related to the topic of IVHM. A couple of them - IP-92 [12] and CIP-105 [13] - are on structural health monitoring (SHM), which we will discuss in the next section, and IP-180 is on AHM [14]. IP-180 is a good example of the industry looking ahead and and dealing with how novel IVHM ideals might be incorporated into the MSG-3 process.

The IMRBPB reviews the CIP and issues it as an IP after it is accepted. During the review process, a CIP could be rejected, or sent back for rework. Once it is accepted, the IP is published and is available to the larger aviation community for review and commentary. The published IP is the primary means by which the IMRBPB recommends changes to the MSG guidance for maintenance. The MSG document is revised every three years by incorporating these accepted IPs. Once the update is codified in the MSG document, any OEM can use it as guidance in the future to incorporate a new process into their maintenance plan. IPs are a very effective way of socializing any change or clarifying positions within the community. A convenient website has been set up by EASA to maintain IPs and CIPs and make them available to the public: https://www.easa.europa.eu/document-library/imrbpb-issue-papers. In this context IP-180 [14] is a particularly good example of the outlined process. It's goal is to modify the existing MSG-3 maintenance logic to incorporate AHM capabilities so that a consistent industry-wide approach to AHM can be established. The authors - who are part of an AHM working group within MPIG - propose an enhanced logic diagram that adds another level of maintenance activities to those already expounded in MSG-3. It is described as:

> *Level 3 - If the system offers AHM capability, a third level decision logic (i.e. Level 3) may be applied. This level enables working groups to assess failure causes covered by AHM capability associated with lubrication and servicing, detecting degradation, and detecting hidden failure.*

While the recommendations from IP-180 have not yet been incorporated into MSG-3 guidance, the plan is to do so in the next revision in 2021.

AC 43-218

Another route by which certification and airworthiness authorities can convey their thinking on a specific topic is through ACs. In anticipation of perceived changes in technology, these organizations often write guidelines proactively to advise the industry and elicit feedback to make their guidance more solid. One such technology area that is of current interest is aircraft health management. Because the FAA has seen a lot of activity related to this topic within the industry and standards-setting organizations like SAE International and ISO, they have decided to publish guidelines for how an applicant should approach the authorities to get approval for a PHM function.

As mentioned in the introduction, the rotorcraft industry has been a pioneer in this methodology with AC 29-2C [6]. In 2003, the FAA introduced miscellaneous guidance MG15 to address the use of HUMS for obtaining maintenance credits, marking a point where authorities saw a need for a guidance document to be issued for PHM technology in this industry. They ultimately added this appendix to the existing AC. Fixed-wing commercial aviation authorities are doing a similar thing now.

The draft AC 43-218 was published in the summer of 2019 and has now been revised based on the feedback received from the community [7]. It is expected that the final version will be published in 2020. The following discussion is based on the review of a pre-publication final draft. The purpose of the document is quite clear [7]:

> *Aircraft health monitoring for maintenance uses onboard sensors, data transmission, and data analysis to provide information regarding aircraft system performance and structural condition. The result is then used to make aircraft airworthiness determinations that provide economic efficiencies while maintaining or enhancing operational safety. This end-to-end process is known as Integrated Aircraft Health Management (IAHM). This AC provides guidance for developing an operator's IAHM program. This AC describes an acceptable means, but not the only means, to comply with the applicable sections of Title 14 of the Code of Federal Regulations (14 CFR). However, if you use the means described in this AC to show compliance, you should follow it in all important respects.*

Interestingly, this uses language very similar to AC 29-2C MG15 [6], the rotorcraft certification document. It defines an IAHM system as an end-to-end process and recommends that applicants take this into consideration when performing an analysis for system approval. The audience for this AC consists of operators and the maintainers of aircraft or all categories, but the guidance applies to design approval holders (DAH) of the IAHM system. When an IAHM system is designed into a new aircraft, the DAH will be the airframer, but for retrofit solutions, the DAH will be the holder of the STC. Like AC 29-2C MG15, the use of commercial off-the-shelf (COTS) systems is discussed in AC 43-218, but not in as much detail. It also acknowledges that ground-based equipment will be part of the IAHM end-to-end monitoring process [7]:

Equipment and infrastructure that is typically used to process and display data collected during aircraft operation. The ultimate use of the processed data will be to make decisions pertaining to some intervention action or provide data to other processing means to make the intervention action determination. Since the ground-based equipment is an important part of the process for determination of intervention actions, its integrity and accuracy requirements should be equivalent to other parts of the IAHM process. In some cases, the ground-based system is provided by the DAH. However, the majority of ground-based systems are comprised of COTS devices that may be required to conform to a minimum performance standard specified by the IAHM DAH when intended for maintenance credit.

Clearly, the use of COTS systems will become increasingly critical in all aviation applications, especially when the applications involve any complex ground-based equipment. But even on board, the use of handheld commercial devices such as tablets will mean that the safety analysis of the entire system will have to deal with COTS equipment. This is not going to be easy; it will bring up interesting issues regarding the testing and verification of such systems.

The draft AC 43-218 lays out, at a high level, the methodology for an applicant to introduce IAHM for a specific aircraft. The authorities wanted to stay away from proprietary terminology or trademarked abbreviations, so they came up with a new term - IAHM - to describe this new functionality, but they acknowledge that IAHM encompasses older systems such as Engine Condition Monitoring, SHM, Aircraft Health Monitoring, etc., some of which might be proprietary to the users of these systems.

This document does not prescribe the methodology for the design of any IVHM system, nor does it describe in detail the steps for achieving its certification. It merely lays out the steps for how a potential DAH would gain approval for such a system, assuming that a technically feasible solution has already been developed and tested. It also specifies what needs to be done if the IAHM system used to obtain maintenance credits itself fails during normal revenue operation. Because this is an AMOC, it suggests that the operator may be able to fall back on an older inspection methodology - the most logical action to take. The only issue would be when the IAHM system is the sole source of data to decide what maintenance action to take; in which case, the DAH will have to specify the AMOC to ensure continued airworthiness, because the failure of the IAHM system would itself be considered critical if the system that it is protecting is critical. The AC discusses not just the steps leading up to certification but also the supporting actions that need to be taken, such as the training of personnel, inclusion of the IAHM system in the minimum equipment list, actions related to the generation, transmission, storage, and security of data related to the IAHM system.

This initiative from the FAA is an acknowledgment of that fact that PHM is an important emerging technology in the aerospace field, and is here to stay. It will influence maintenance programs and will be used for maintenance credits in the future. This AC is a clear indication that automated monitoring and health management systems are not to be feared and that the authorities are open to applicants approaching them for approval. However, this does not obviate the need to prove that IVHM systems are technically beneficial and economically viable, and as safe as the scheduled maintenance actions that they are replacing.

Recommendations

No new maintenance procedure will be approved unless it goes through the approval steps that involve the certification and airworthiness authorities. This is true for paradigm-shifting technologies involved with PHM, because the basic goal is to eliminate all regular inspections and move to a CBM philosophy. This brings up a host of issues that will need to be addressed before these technologies start to be routinely used in commercial aviation.

- Standards development organizations can take a leading role in generating advocacy for IVHM system by publishing guidance material related to specific technologies, as well as general principles. This is already happening to an extent within SAE International committees like E-32 (Propulsion Health Management) and AISC-SHM (Aircraft Industry Steering Group for Structural Health Management). These committees have already published some documents that address the issue of maintenance credits directly. SAE AISC-SHM published ARP6461 in 2013 and SAE E-32 published ARP5987 in 2018, both present guidance on how IVHM technologies can be used to obtain maintenance credits. In the future these committees, as well as the HM-1 (Integrated Vehicle Health Management) committee and the IVHM steering group will be working on advancing this cause.

- The issue of data ownership will have to be addressed. All stakeholders will have to realize that it is in their interest to collaborate. There might be a role for neutral standards development organizations in facilitating this noncompetitive collaboration. This would be very similar to the MSG consortium that came together 50 years ago to develop maintenance standards. In addition to data ownership, there is the issue of data sharing, where different stakeholders are willing to share proprietary data to help resolve a maintenance issue. This is easier to envisage in theory, but in practice would require negotiations and well-crafted contracts. In addition to the contractual issues, this will bring up technical issues such as data security and access, which will have to be solved.

- FAA advisory circular AC 43-218 - to be published in 2020 - will lay out a set of overall guidelines for IAHM systems. To make these more useful, additional guidance for individual technologies will have to be worked on, possibly standards development organizations. Once the first few systems are certified using these guidelines, the path will become clearer and easier for subsequent applicants.

- Since one of the ways PHM functionality can make its way into standard maintenance practice is through a revision of the MSG-3 guidelines, a good strategy is for someone to author an IP on this subject and present it to the MPIG and the IMRBPB. These MSG-3 guidelines are revised every three years and the next revision is scheduled to be published in 2021. More than one IP may have to be written to outline the key elements of IVHM systems, or to describe specific technologies that can help modify the procedures, such as brake monitoring described in this section. IP-180 is just such a document. In addition to recommending changes to MSG guidance it also lists some examples of maintenance tasks that can be modified using AHM capabilities using the enhanced decision logic that is presented in the IP.

- Closely related to this is the development of standards. An aerospace information report (AIR) describes any aspect of a specific maintenance procedure and how that might be changed with the existence of novel IVHM systems. Such a document might be a good avenue to move the conversation forward and support applicants and authorities during a process review (e.g., during an MRB meeting). While AIRs may be specific to SAE International, there are sections of the aviation landscape that are not covered by some of the committees, such as power electronics. That could be taken up by other standards development organizations with more relevant knowledge about the subject.

The Rotorcraft Experience

It is interesting that, in the application of IVHM systems to obtain credits of any kind, the rotorcraft industry is ahead of its commercial fixed-wing aviation counterpart. This is partly because different certification rules govern different groups in the industry, but mainly because the consequences of failures with helicopters is not as catastrophic as they are with airliners. This has made the safety of large transport category aircraft of paramount importance during system certification, sometimes to the detriment of innovation. However, there are lessons to be learned from the rotorcraft world that can help move the fixed-wing community towards a more reasonable use of IVHM systems for supporting maintenance. This section of the report is influenced heavily by

discussions with Mark Davis, recently retired from Sikorsky Aircraft Corporation (SAC).

One of the most remarkable documents that the FAA has produced related to this topic is MG15, which was published as an appendix to advisory circular AC 29 [6]. This MG was incorporated into the advisory circular document in 2003 and is titled "AC 29 MG15: Airworthiness Approval of Rotorcraft Health Usage Monitoring System (HUMS)." The history of HUMS in the commercial rotorcraft industry is rather checkered. It started off in the 1990s with vibration monitoring systems being installed on helicopters operating in the North Sea, and the United Kingdom Civil Aeronautics Administration issuing an airworthiness directive 001-05-99 in 1999, making its use on UK-registered passenger helicopters mandatory. US Army helicopters were also retrofitted with HUMS units (called Data Source Collectors) around the same time. In both cases, significant improvements in operational safety were observed. In addition, according to the US Army, cost savings of $112 million over a 26-month period were determined after the installation of these units [31]. A good survey paper on HUMS usage for both military and commercial applications is the one presented at the AHS conference in 2017 [33].

MG15 lays out some of the requirements for a HUMS function to get FAA airworthiness approval. Remarkably, it spells out not just the installation aspects but also the end-to-end process for the use of HUMS to obtain credits in a manner very similar to AC 43-218. For example, while defining this concept of an end-to-end process, it says:

> The term "end-to-end" as used in the text is intended to address the boundaries of the Health Usage Monitoring System (HUMS) application and the effect on the rotorcraft. As the term implies, the boundaries are the starting point that corresponds with the airborne data acquisition to the result that is meaningful in relation to the defined credit without further significant processing. In the case where credit is sought, the result must arise from the controlled HUMS process containing the three basic requirements for certification as follows:
>
> i. *Equipment installation/qualification (both airborne and ground),*
>
> ii. *Credit validation activities, and*
>
> iii. *Instructions for Continued Airworthiness (ICA) activities.*

This is one of the first places where the concept of maintenance credit was defined. According to this AC, a credit can be defined succinctly as

> Credit: To give approval to a HUMS application that adds to, replaces, or intervenes in industry accepted maintenance practices or flight operations.

The advisory circular clearly points out that when the intended HUMS feature is being applied to obtain credit,

FIGURE 5. Flowchart for application of HUMS-specific software with a ground base that uses COTS software for operational software [6].

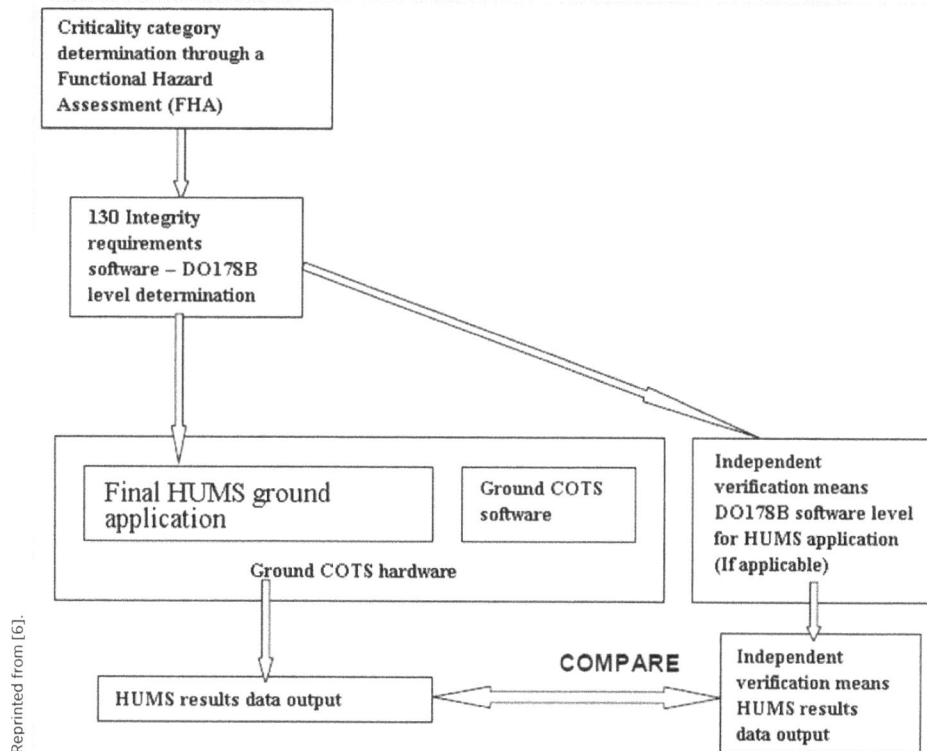

Reprinted from [6].

the criticality of the system must be determined from the "end-to-end" perspective, including airborne and ground-based elements. This means that if there are COTS elements in this data chain, they should be included in the criticality analysis. It also advises that if the applicant is intending to use the system in the future for a higher criticality function, it might be a good idea to certify the system at the higher level; otherwise a reevaluation will be required. Finally, it is important to note that MG15 does not address systems where the consequence of failure is catastrophic. While this poses a restriction on the application of MG15, because of the experience gained by companies during the application of MG15 to their systems, we should be in a better position to evaluate what needs to be done for higher criticality systems in the future.

For now, partly because of this restriction, MG15 has been able to publish guidance on the use of COTS software. The issue with the use of COTS software systems is obvious from a DO178 perspective. All airborne software code is supposed to be written from scratch, so that it can be tested thoroughly against all contingencies. Also all the source code must be available so that it can be reviewed by a team other than the one writing it. With a COTS system this is not possible. The source code will often not be available, and the history of the development would have been long forgotten. The only thing available to test would be a "black box."

MG15 recognizes this restriction and allows COTS software to be qualified in several ways, such as with historic operational data, use of dissimilar computers, physical inspections, etc., but always backed up by independent verification. This independent verification can be stopped once confidence in the COTS software and hardware is established over a period of time. MG15 makes very clear that all these are for ground-based equipment only, and that airborne systems must follow DO178/DO254 criteria for the development at the appropriate assurance levels. Figure 5 shows the flowchart for mitigation action.

By adopting the guidelines in AC 29-2C, the rotorcraft industry has managed to make some remarkable strides with respect to applying IVHM systems to obtain maintenance credits for critical components. While there are several examples, the one presented here has been documented in an FAA report and conference papers [2, 3]. The Sikorsky S-92 is a modern helicopter that was introduced in 1998 and is used in commercial transport operations, including in the oil and gas sector.

As stated in the introduction, one of the central goals of usage-based maintenance (UBM) is to change the maintenance paradigm from one that is currently flight-hour or schedule-based to one that is based on usage and loads monitored in the operational rotorcraft fleet. Life-Limited Parts (LLPs) are currently retired based on the number of flight hours flown regardless of whether the aircraft is flown

FIGURE 6.   **HUMS installation and maintenance credit validation flowchart.**

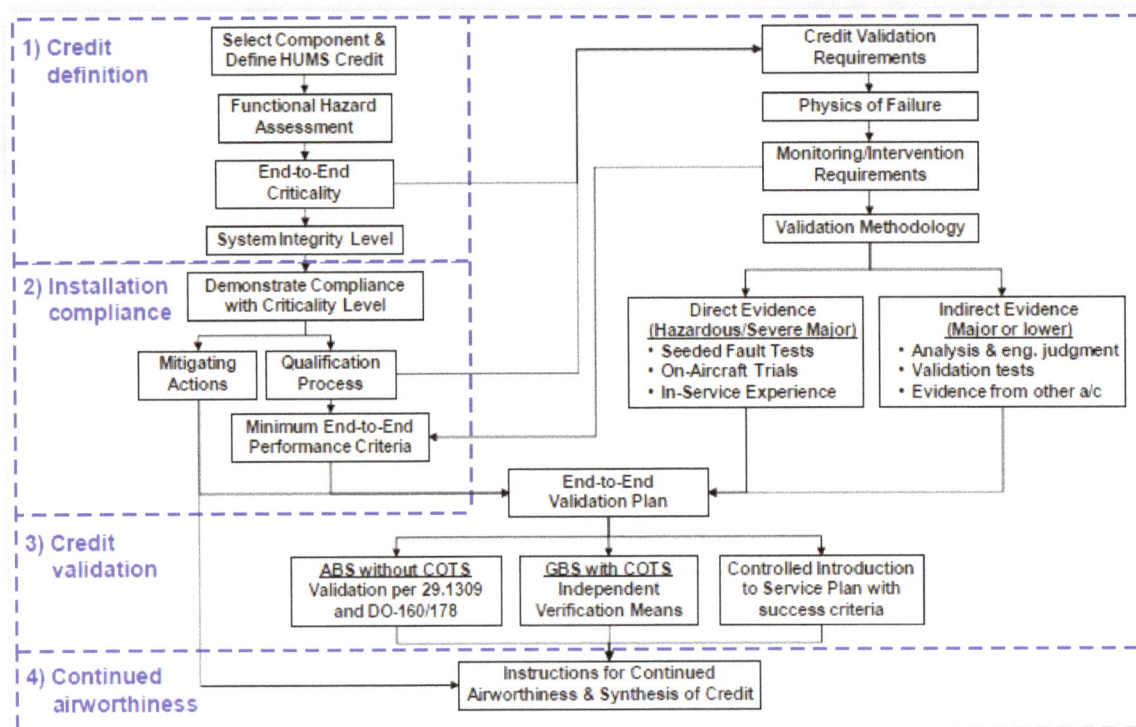

benignly (e.g., VIP transport) or flown much more aggressively (e.g., cargo transport with many ground-air-ground cycles). A certified UBM process that accurately monitors usage and loads would enable the attainment of retirement time credits (or debits) for LLPs based upon measured aircraft configuration, usage, and loads rather than flight hours and underlying conservative design assumptions. SAC is making steady progress in achieving this long sought-after goal.

The way SAC engineers go about achieving this goal for the main rotor blade of the S-92 is by downloading data gathered by the HUMS box, analyzing it, characterizing the load regimes that the helicopter has flown, and estimating how much of design life has been expended. Essentially, this has allowed SAC to determine that the main rotor blade retirement time can be extended from 22,000 FH to more than 30,000 FH. This is a staggering amount of benefit for the operators and could only be substantiated because of the analysis that was done with more than 700,000 hours of data gathered by the HUMS. Normal life limits for the main rotor blade is 22,000 FH with maximum gross weight of 26,500 lbs, and this drops dramatically to 14,000 FH when the maximum gross weight goes to 27,700 lb. By analyzing data from the entire fleet, SAC engineers were able to positively determine that the original design life limits were too conservative because the assumed aircraft loading was too high. Redoing the calculations, without changing any of the design fatigue calculations, allowed them to increase the safe life limit to more than 30,000 FH. This procedure followed

the AC 29-2C process with the HUMS unit on the S-92 as illustrated in Figure 6. The company has applied to the FAA for this credit, which is still under review.

While this is a clever use of HUMS and UBM, it is only a part-level credit (i.e., the credit is extended for the entire part family). True PHM would require this to be done at the serial-number level where individual serialized parts are tracked, and different life limits are assigned based on usage. Sikorsky has done this as well for the main rotor hub, but no detailed documentation is available. In the paper describing the main rotor blade life extension, the authors point out that

> *In 2012, SAC obtained FAA approval for life extensions to S-92® main rotor (MR) hubs based on the actual ground-air-ground (GAG) cycles recorded for individual hub components in the field. The methodology utilized HUMS data combined with a usage monitor reliability factor (UMRF) (Ref. 7) to determine a one-time life adjustment for a particular MR hub serial number.*

The usage monitor reliability factor is described in a paper by Adams and Zhao [1] and refers to a method for ensuring that operational reliability is maintained by adjusting the counts of flight regimes while tracking usage over different regimes to account for the full range of usage scenarios. This approval is for individual serially numbered parts, which means that this counts as an example of a true

CBM application in aviation. While this is better than the part-based UBM credit, it is still not as beneficial as it could really be, because it is a one-time life extension and not a continuous process. The hope is to extend this to a continuous process whereby an automated system continues to accumulate credits as it analyses rotor hub usage. While SAC has been a pioneer in taking advantage of MG15 guidance to deliver HUMS-based benefits for its customers, it is not the only company to do so. Others are following suit, though they are not ready to discuss it in public just yet.

Recommendations

The point of presenting all these examples and the published AC is to show how the rotorcraft industry has taken the basic ideas of tracking actual usage and employed it to obtain maintenance credits. This allows the OEMs to deliver actual quantifiable benefits their customers. There is nothing to prevent the commercial fixed-wing aircraft industry from doing the same.

- The AC that the FAA is already in the process of publishing (AC 43-218) seeks to give some high-level guidance on the certification of health management systems. It is not as detailed as the miscellaneous guidance (AC 29-2C MG15) that is available to the helicopter industry. But it is a very important first step and demonstrates that FAA is very keen on seeing these systems be used in Part 25 aircraft. However, this is only a beginning. Something equivalent to MG15 for commercial transport aircraft is needed. While it is not going to be straightforward, the fact that the industry is starting to move in this direction is a good thing. The new AC may have to tackle systems at higher criticality levels as well. Issue paper IP-180 addresses some of these issues, as described in the previously section, by modifying the maintenance decision logic to add AHM. But it will be a while before its recommendations will translate into guidance for aircraft maintenance.

- Engine and airframe OEMs should study the helicopter industry to determine how their example can be replicated in the fixed-wing world. Yes, the safety considerations can be a little more stringent because of the criticality levels involved, but if the process is followed properly, then there is no reason why IVHM systems that mitigate higher criticality failures cannot be designed and certified. Some of the examples that might constitute a beginning are given in the next section.

- The use of COTS equipment (hardware and software) in health management system is inevitable, and the sooner the industry figures out how to deal with their use the better. Short of demanding the source code and testing every line, designers will have to come up with advanced testing methodologies that ensure that the

equipment cannot malfunction in any possible scenario, within, of course, the parameters of uncertainty that are laid out by the recommended practices acceptable to the industry. Some of this has been tackled by the authorities for ground-based systems, but a far more detailed analysis must be undertaken for airborne systems to ensure that the current levels of safety are maintained.

Specific IVHM Use Cases and Issues in Commercial Aviation

The standards for maintenance have changed over the years and some tangible progress has been made in the application of IVHM technologies. On-condition maintenance was only the beginning of this trend, and the industry is moving towards more and more condition-based practices. These PHM or CBM technologies support a general class of maintenance practices that aim to schedule maintenance based on the actual condition of the part. In this report, this is taken as a catchall phrase for all IVHM practices that do not follow strict schedule-maintenance procedures. For some practitioners, there is a difference between on-condition maintenance and CBM. The former is where maintenance actions are triggered by condition indicators or regular inspections. The latter is completely automated where no inspections are needed for the condition to be estimated. The brake-monitoring example we described in a previous section is an example of what is possible when we replace a scheduled inspection task with an automated condition monitoring feature. While it is a hypothetical example, the underlying technology makes the example realistic. There are several examples of PHM-based maintenance procedures which are either already being practiced or can be practiced without too many modifications. We will describe some of them here. The idea is that by studying these we can hope to gain insight into what needs to be done to make these practices more widespread in the industry. This section has been informed by discussions with Bal Annigeri, David Piotrowski, Ian Jennions, Pierre-Charles Rolland, Rhonda Walthall, and Guilherme Torres.

It should be noted that IP-180 [14] also presents examples of maintenance functions that can be modified via the application of AHM capabilities. In fact, it lists these in its Appendix and walks through the modified decision logic diagram to point out how the logic will flow while each function is considered. The maintenance functions analyzed in detail in IP-180 are

- Removal of inertial units for calibration

- Visual check for brake wear

- Functional check of the gross internal leakage of the hydraulic system

FIGURE 7. **Throttle position/correlated damage during different flight segments.**

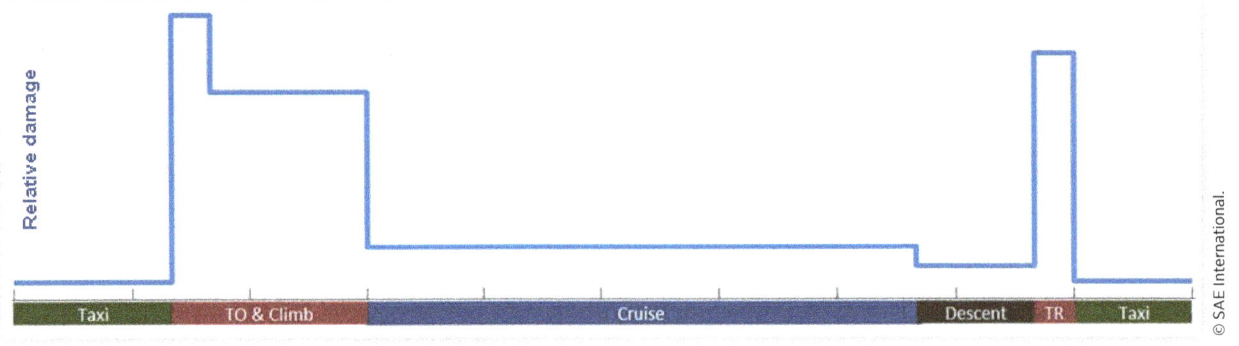

- Operational check of hydraulic systems reservoir quantity indicator
- Restoration of service pressure regulator filter
- Oil change
- Replacement of recirculation air filters

All these examples are studied in detail and suggestions made as to how AHM can help change the schedules associated with these tasks. It is a very insightful document and should be studied in conjunction with this section to understand how IVHM systems can be introduced into commercial aviation.

In addition to this issue paper, examples of the development and use of health management functions within the aero-propulsion field can be found in a collection of papers edited by this author [23].

Usage-Based Lifing

One of the most important classes of aircraft parts are Class A parts, also known as LLP. These parts tend to be expensive and are mainly found in high-stress and high-temperature areas like the landing gear and jet engines. We will discuss engine rotors in this section, but the same idea would be applicable to any LLP. In addition to being expensive parts, rotors are also expensive to replace, because they entail an entire engine teardown. Therefore, it is important that no useful life is wasted by replacing them sooner than needed. Being able to estimate remaining useful life (RUL) accurately would go a long way in ensuring that this is accomplished.

Parts that have life limits can only fly for a given number of hours or cycles because cyclic loading at elevated temperatures will force them to reach their fatigue limits. Cracks within the structure will first initiate and then, as the part is loaded, it will gradually grow to be larger than a safe size. At this point the part would have reached its life limit and would need to be replaced. Doing a scheduled nondestructive evaluation of each part during operation is prohibitively expensive, so the process of assessing life expended is currently based on analytical methods. During the design of these parts, standard fatigue lifing theory and selective material and endurance testing establish life limits under conservative conditions. For the most part, this limit is used regardless of actual operating conditions which, for most flights, would be less stressful than the worst-case conditions assumed for design. Figure 7 shows the relative damage that a flight cycle imposes on an aircraft engine rotor (where TO refers to takeoff and TR refers to thrust-reverse operation). This is correlated strongly with the thrust command, but thrust levels will be higher or lower depending on loading, ambient conditions, and other factors as well.

With standard life tracking, each flight cycle will result in a debit of one flight cycle from the life of the rotor. Let us assume that the total life of the rotor is N cycles. Given such a conservative methodology of debiting life, the total useful (calendar) life of the rotor will be Nf hours, if f is the average duration of a flight. This is because the LLP will be replaced the moment N cycles are complete.

UBL works by extending the useful age of an LLP by taking into consideration the actual conditions under which the aircraft is flown. For example, if the passenger count is lower or the ambient condition is colder, the amount of thrust needed to takeoff, climb, and cruise would be lower. If this were taken into consideration (as in Figure 8), this could result in an extension of the life of the part which could be significant, depending on how loaded each flight happens to be. For example, if UBL were being used to count life for this flight, the actual damage sustained by the rotors would be a fraction of the assumed worst-case damage and, hence, would be some number less than one. Counting in the same manner over the life of the engine will lead to an average credit given to the rotor over the course of its life which would also be less than one; let us assume for the sake of this example that it is d, then the number of actual flights that the rotor can complete with the same life limit would be N/d instead of N. Given the same flight length f, the actual life of the part would be Nf/d, which could be significantly larger than Nf depending on how much smaller d is compared to 1. This is beneficial to the operator because of the high cost of these LLPs. It is in the interest of the operator to ensure maximum usage out of these parts without compromising or violating any original design safety limits.

FIGURE 8. Derated throttle position/correlated damage during different flight segments.

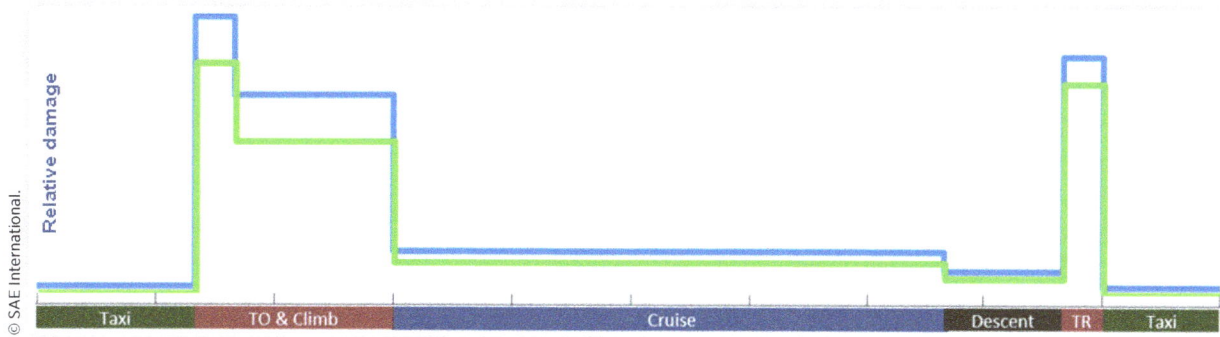

A more detailed explanation of UBL can be found in the conference paper presented at the Aerotech in Toulouse in 2011 [24]. This was a summary of an earlier version of an aerospace recommended practice that the SAE E-32 committee was writing, which included several examples of systems that could be used to get maintenance credits for aircraft engines systems.

In fact, the SAE E-32 committee for Propulsion Health Management has been working for many years now on developing guidance on how maintenance credits can be obtained for specific IVHM technologies. The latest relevant document that SAE E-32 published on this topic is ARP5987 [29]. This recommended practice is a great beginning, but it does not go far enough in its journey. To go the whole way, we need a concentrated effort to advance the processes by which maintenance credits can be obtained.

Derated Operations The fact that LLPs are expensive and that there are benefits in making them last longer is not lost on engine OEMs. Even though they make more money by selling LLP spares, it is not a feature that endears them to their customers. One of the ways they have sought to help their customers is by introducing the concept of derated operation.

The engine companies publish derate schedules in the engine manuals. A good description of this is available in SAE AIR1872B published by E-32 [26]. While this document has officially been cancelled because of a lack of interest from the community, it is nevertheless a good reference for understanding derated operations. If we are to truly implement UBL as a PHM function for maintenance credits, this document would need to be resurrected, reviewed, revised, and reissued.

The life of an LLP is limited by a number of factors:

- Low-cycle fatigue (LCF)
- High-cycle fatigue (HCF)
- Thermal fatigue
- Creep

There are several other nondeterministic causes for failure (e.g., foreign object damage, erosion and corrosion,

mishandling during maintenance), but the physical failure mechanisms listed previously are subject to physics-based analysis and are much more systematically predictable. (More random factors need to be monitored on an ongoing basis to ensure that they do not contribute to limiting component life.) Life limits are based on operations at a given worst-case scenario. An LCF cycle is the easiest to measure because it consists of one flight cycle. HCF is much more difficult to measure and assess and, therefore, does not play a significant role in LLP lifing calculations. Rapid thermal changes can create adverse conditions for critical parts, and unusual occurrences not accounted for in a normal LCF cycle are tracked. Creep, like thermal fatigue, is not a common occurrence for engines except if they are operated for long periods of time (e.g., cruise) at elevated power levels, which does not happen often.

Because the takeoff is considered the most damaging maneuver for an LLP in an engine, the takeoff report is used to assess the derate factor. Reverse thrust is generally just accounted for as a discrete value based on its usage during landing. If the pilot considers ambient conditions and load factors and determines that he or she can take off at a lower power setting, then the engine manuals allow for a derate factor to be applied to LCF life. This works just like the factor d in the previous discussion. The takeoff report will record this lower thrust operation, and the maintenance personnel can use this to discount the damage for that particular flight. If carried out systematically, these credits can be integrated over the lifetime of the part, to extend life. Appendix C in AIR1872B gives a great example of the use of derate factors for the Rolls Royce RB211 engine. Unfortunately, because of all the work that is involved in calculating the derate, it is often not carried out, and the benefits are lost. Today, the entire process can be automated and the necessary calculations done by computer.

Taking this automation idea even further, a true UBL system would not require any post calculation. The entire system would be automated and qualified at the appropriate design assurance level. This would require more work during the design stage to ensure that all necessary safety analyses are performed as specified in ARP4754 [27] and ARP4761

[28] to ensure reliability. Details of a possible system are given in the Aerotech paper [24]. One of the key challenges to accomplishing this is ensuring that the data integrity for UBL calculations is maintained during its transmission from the aircraft, through telecommunication channels, and to the ground-based system that is used for the analysis. Another key part is the use of COTS hardware and software involved in the ground-based equipment. In the cited paper [24], some simple strategies (e.g., circular redundancy checks) are recommended for maintaining data integrity during transmission. The discussion in the previous section on verifying COTS equipment is immediately applicable to this example as well. The cited paper also describes how the algorithms can be validated so that they can be used for making major maintenance decisions, especially those that allow LLPs to stay on wing longer than older strict life counting would have allowed.

Oil Debris Monitoring

The most common practice today for monitoring metallic chips in engines is based on capturing particles on magnetic surfaces and examining them visually. However, a few advanced systems have been installed on modern aircraft that use changes in electrical signals to detect the presence of metallic debris in oil. They are much more common in military and the stationary gas turbines world than in aviation jet engines. However, some more modern commercial jets have advanced ODM systems installed.

The system from Eaton Aerospace (originally Tedeco Industrial Systems) is installed in the GE90 and the GP7000, among other engines. This system replaces a magnetic chip detector (MCD) by incorporating an inductive coil that can count ferromagnetic chips as they get trapped by the magnet. To quote the company literature on this system [5]:

> *The heart of the system is a magnetic, inductive particle sensor. It captures and retains ferromagnetic debris particles and provides an output signal for each capture. Eaton's Tedeco brand markets this sensor under the trademark QDM® (quantitative debris monitor).*

There are in-line systems as well, such as the one from Gastops, that also work on the inductive principle but are able to detect ferrous and nonferrous particles. These have been used on military applications and on land-based gas turbines [11]. The precise technology is not relevant to our discussion here, only the PHM function related to these sensors is. Let us assume that the reliability and the accuracy of these sensors is good enough to eliminate any regular chip detector inspections, in favor of condition-based inspections.

Current maintenance practice for engine oil inspections is to inspect MCDs after a fixed number of flights. MCDs are simply plugs with magnetic heads that stick into the oil stream (generally near the lines that scavenge oil from the bearing compartments) and capture ferrous particles. They are packaged in such a manner than they can be removed for inspection without the oil leaking out. The inspection regimen for MCDs requires the maintainer to visually examine the chips and assess whether this represents a cause for concern. Because of the nature of the inspection this can be subjective. In addition to the visual inspection, the maintenance procedure requires chips to be removed from the tips and sent to a well-equipped laboratory for further spectroscopic analysis. This allows trained technicians to do a thorough material assay and thereby determine the extent of damage. The idea with the ODM system would be its use in an in-situ analysis system to automatically assess the amount of debris as well as perform an oil analysis. With this information, a decision can be made to either extend the inspection interval or eliminate all regular inspections altogether, in favor of condition monitoring.

In this future scenario, the engine OEM would work with the ODM supplier to determine the overall effectiveness of the solution. The whole point of this exercise would be assuring all the stakeholders that an automated system that replaces human inspections would be just as effective in finding issues with the lubrication system. The objective is to make sure that system safety is not compromised. These studies could include tests to determine the capture efficiency of the system (i.e., what fraction of existing chips could the ODM detect and what is its performance with respect to chip sizes).

For any diagnostic system, the two key metrics are the detection rate and the false-alarm rate. In fact, a graph plotting one against the other is often used as a measure of the effectiveness of a diagnostic; it is called the receiver operating characteristic. The area under the receiver operating characteristic for a given threshold value is a comprehensive metric for the detection algorithm. The name is derived from the origins of this technique (i.e., detection of targets using radar).

While these are some key metrics for any detection system, there are many more useful metrics for IVHM systems [32]. The point is that the OEM would work with the supplier of the equipment to make sure that the fundamental metrics associated with the device satisfy the overall safety criteria for the engine.

There are systematic ways to determine safety levels for aircraft systems. At the aircraft level, the goal is to prevent a catastrophic event resulting in loss of life. The reliability of everything within the airborne system is assessed against this eventuality. Reliability is an important metric because it is understood that a 100% level of safety is nearly impossible to achieve in any engineering system. In the design and analysis of aerospace systems, the definition of a catastrophic event is something that has a probability of continuing safe flight lower than one-in-a-billion per FH [27, 28]. This is the probability of occurrence for a failure that is associated with a catastrophic failure condition (see Figure 2).

One way to mitigate against a potentially catastrophic failure is to redesign the system so that the consequence of a failure is not as bad. An equally effective method is to add redundancy into the system which, in the simplest case, will make the overall system much less susceptible to a single failure. That is the whole purpose of going through a detailed failure modes and effects analysis, fault tree analysis,

FIGURE 9. SHM in the context of vehicle heath management.

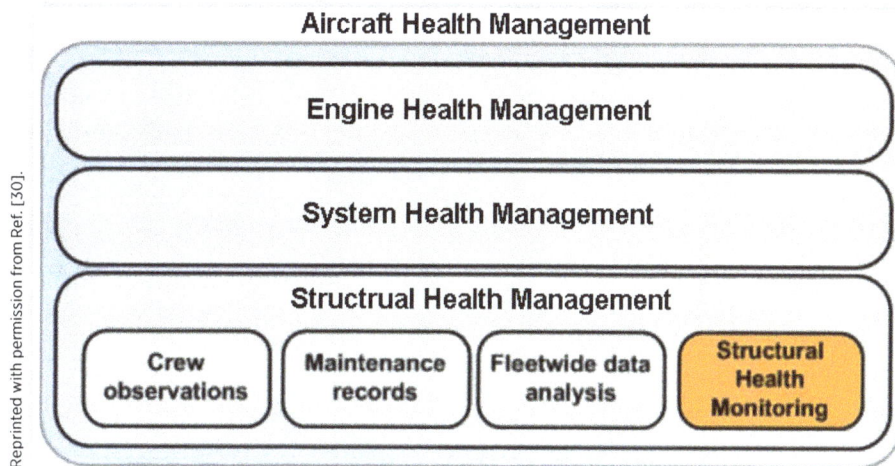

Reprinted with permission from Ref. [30].

functional hazard analysis, common mode analysis, etc. All these techniques are spelled out clearly in ARP4761 [28] and are essential for ensuring that the overall airborne system is safe. The goal of using the ODM system to replace or extend MCD inspections will depend on its ability to provide a reliable alternative to the latter.

From a practical point of view, the most probable scenario would be to let the ODM operate in parallel with regular or extended inspections to test out the efficacy of the new system. Once it has been shown to be effective for a statistically significant number of samples, it would be clear that using the new maintenance practice would still be as safe as the standard one. This has been done with existing engines such as the GE90 to lengthen the inspection intervals and save on line maintenance labor. The next step would be to make the system completely condition based. Debris in oil is only one form of contamination and a measure of oil degradation. Physical properties such as oil quantity and chemical properties such as viscosity and oxidation are additional required measures that can be used to assess when oil needs to be changed. Some military aircraft engines already have automated oxidation and viscosity measuring sensors installed in the oil, but this technology has not yet found its way into commercial aircraft engines. To be able to truly automate oil monitoring, these advanced measurement techniques will have to be incorporated.

Structural Health Management

In recent years, major PHM developments have been made in nondestructive testing (NDT) and SHM. These are related technologies, where the structural health of the component is assessed in situ without having to destroy it. We will discuss SHM by using the example of a system that Delta Air Lines

has successfully used to monitor a specific part in their aircraft. This is to obviate the need for constant monitoring.

Traditionally, NDT methods have been employed during major airplane checks to assess structural damage. When these techniques are enhanced with in-situ sensing techniques that can continuously monitor aircraft structures and report on their health, it is called SHM. Two papers from the SAE International Journal of Aerospace give a good overview of the subject and how SHM relates to MSG-3 [10, 16]. These papers were written by members of the AISC-SHM, which was set up by SAE as a technical committee in 2008. This committee has been working diligently to develop guidance on how SHM elements can be incorporated into commercial and military aviation. They have produced an aerospace recommended practice (ARP6461) that presents guidance for how SHM can be incorporated into standard MSG-3 practices [30]. Figure 9 (from ARP6461) shows how SHM is structured within the context of vehicle health management.

There was an opportunity during the development of the Boeing 787 Dreamliner to incorporate more SHM techniques into its maintenance practices, but this was an opportunity missed. The AISC-SHM was formed to ensure that the industry does not repeat this in the future and that more SHM elements are included in any new development, be it commercial or military. Aspects of SHM have been introduced in the 2009 revision of MSG-3 (MSG-3 2009.1) that dealt with scheduled-SHM techniques. Following this guidance, OEMs can apply to replace some inspection procedures with scheduled-SHM techniques. Scheduled-SHM tasks are essentially the same as NDT tasks except that the system used is closer to what would be used in a more automated system. In contrast to their scheduled counterpart, automated-SHM would employ systems that would continuously monitor components within the aircraft. This is the next step in the evolution.

One automated-SHM technique that has been used by Delta is called Comparative Vacuum Monitoring (CVM) [19].

This was funded by the FAA to move SHM from a prototype system to a routine maintenance procedure, and the goal of this project was to develop guidance for the industry concerning a specific SHM technique. In this technique, a flexible film with alternating vacuum and atmospheric-air channels is applied to the surface where a structural failure is expected. When a crack develops, the vacuum in the channels above the crack cannot be maintained, and this change in pressure indicates not just the presence, but also the location of the crack. The incentive for Delta is that it will allow them to replace visual inspections of the suspect areas with continuously operating CVM systems, which should reduce cost. One of the parts they are targeting is the Boeing 737NG front spar shear fitting in the center wing box, where some of which were starting to develop cracks at a relatively low cycle count.

The installation procedure is not easy and ten such installations were required for each aircraft to cover the entire span of the wing box (see Figure 10). The system was validated by visually inspecting the parts that were being monitored by the CVM system. Delta TechOps is in the process of developing a standard maintenance procedure where the CVM system will replace regular visual inspections and provide advanced warning of any cracks. The system, developed by an Australian company, Structural Monitoring Systems, plc, seems to be the first systematic application of automated-SHM in commercial aviation. When certified, this could be adopted by Boeing for their entire fleet, which would mean a change to the MPD. Regardless of how it makes its way to the maintenance procedures, lessons learned from this exercise will go a long way in helping change guidance on such techniques for future applications.

Condition-Based Maintenance SAE International's G-11 Maintenance committee began working on a descriptive document on CBM in the early 2010s which caught the attention of personnel from the US Department of Defense. The idea was to look at all the various definitions of CBM that were extant and provide an information report that would provide a definitive description of the field and its use, specifically within the military. Even though there is a rich history of the practice of CBM within the aerospace industry (e.g., the collection of papers edited by this author [22]), there has always been a lack of precision in the definitions of the concepts within the field. The introduction of CBM-Plus (CBM+) by the DoD in the early 2000s further confused the issue. The goal of G-11M was to publish a simple introduction to the area. However, for a number of reasons, little progress was made until 2018, when the document, now designated JA1013, was transferred to the SAE HM-1 (IVHM) committee and is now being developed as a joint aerospace and automotive document.

This will be an important reference piece because it will not only lay out the principles of CBM in a comprehensive manner, but it will also link the subject to MSG processes. While many studies have been done in the past on this topic, this will be one of the few documents to explicitly lay out the process by which CBM systems would make their way into the commercial aviation world with proper approvals. Members of the reconstituted team come from several leading aerospace organizations, including Airbus and the DoD. In fact, the Airbus person on the team has been working on advancing CBM-related practices within the airframer's maintenance organization. Having actual practitioners on the team is of great help in making sure that the recommendations the team comes up with are practical and useful to the industry. A key concept associated with CBM is the P-F interval, which is particularly illustrative of the criteria needed for CBM to be successful [15]; the principals of which

FIGURE 10.   **CVM device installed on the wing box fitting.**

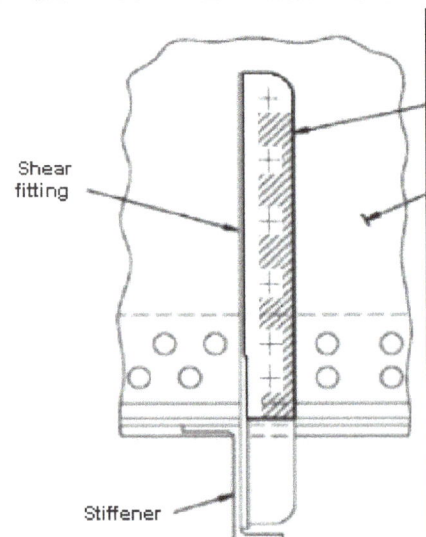

Shear fitting

Stiffener

FIGURE 11. P-F interval explained.

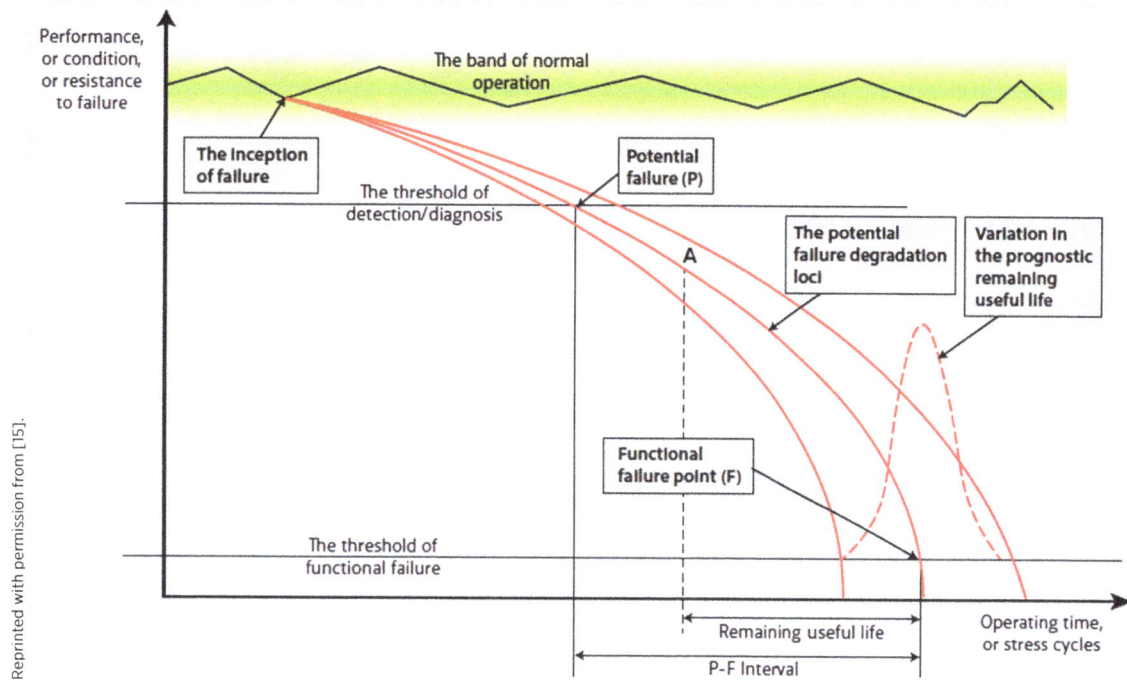

have been detailed by Ian Jennions in *Integrated Vehicle Health Management: Perspectives* on an Emerging Field and are shown in Figure 11. This concept was first put forth during the development of RCM and has now been greatly enhanced by subsequent researchers.

The green band in Figure 11 represents normal operation. The assumption here is that once degradation starts, system performance will steadily degrade, staying within the band of uncertainty. Any real-world system will exhibit variation in future performance, the extent of which will depend on external and internal factors. Estimating prognostic behavior (i.e., estimating RUL) is inherently difficult because of these unknown factors. To be safe, this variation will need to be factored in when making maintenance decisions. In the cited reference, these points are discussed in greater detail. The consequence of failure is explained along with an explanation of the degradation process. The main point that the paper makes is that it is only with systems that exhibit some form of an observable degradation that CBM can be employed safely and reliably. Two thresholds are depicted in Figure 11, the first indicates the threshold of detecting a failure or the point of unacceptable degradation and the second indicates the threshold for declaring the system to be functionally failed.

The choice of thresholds is up to the designer so that the risk of failure is balanced against the desire to use up as much of useful life as possible. Uncertainty in assessing the future of the degradation curve is a critical parameter that factors into the calculation of these thresholds. The larger the uncertainty, the higher the thresholds - or the sooner the part is tagged for maintenance or removal.

As outlined in the first section, one of the ways of getting CBM or PHM functions on the aircraft is by modifying the MSG-3 guidance. However, the conditions currently in this guidance are conservative, so there needs to be a way of working towards a more condition-based approach. This could be based on modifying inspection intervals related to the P-F interval according to a recently written internal issue paper from Airbus [25]. The idea is to use the concept of the P-F interval to develop a set of inspection protocols after the system can detect the failure. In case of structural elements, this would be the point where cracks attain a certain size which can be detected (point P in Figure 11). In this issue paper, the author argues for a safety factor that can be used to shorten the inspection intervals as the system progresses beyond point P and towards the point of functional failure (point F in Figure 11). This allows the system to be more carefully monitored as it operates in a region of higher risk, but still within the bounds of safety, as originally designed. According to the issue paper, this would ensure that the system follows acceptable practices for continued airworthiness and would serve as a bridge from current practices to a more condition-based approach.

Recommendations

In this section, we have looked at a few leading examples of where IVHM has already been applied to maintenance procedures, and we have discussed what these techniques might hold in the future.

- One of the most expensive parts in an engine is an LLP, and maximizing its useful life is a worthy goal. Derated operations get us part of the way there, but to get the most benefit, UBL is necessary. Standards development organizations can play a key role in this by writing guidance documents or even information reports outlining best practices in this area. Documents, such as AIR1872G, can help to educate system designers about IVHM techniques so that concepts such as UBL do not seem too futuristic. These documents will also help socialize these concepts with regulators and make them more receptive to more analytical methods for estimating critical parameters such as RUL. A rewrite of AIR1872 to bring it up to modern standards should be fairly straightforward. The Aerotech paper from 2011 [24] could be used as guidance. UBL is also an appropriate topic for a CIP to be discussed at a future IMRBPB meeting.

- The ODM systems that are already installed on airborne and ground-based applications are producing a lot of data. If this data could be reviewed along with knowledge about the actual condition of the equipment from tear downs, the two pieces of information together could provide insight into determining if existing systems can provide full coverage of oil condition. In other words, could MCD inspections be completely eliminated? Going beyond that, is it possible to eliminate all oil monitoring by incorporating other advanced sensors that could measure additional parameters, such as oil quantity, oil viscosity, and oxidation levels? Incorporating such sensors would be expensive, but it could save a lot of money in the long run by reducing the number of oil analyses in the course of oil life.

- Scheduled-SHM has been part of the MSG-3 guidance since 2009. The AISC-SHM committee should be working towards including automated-SHM in the next revision of the MSG-3, which will be published in 2021. The CVM work that is being doing by Delta could be used as an example to develop some basic guidance. In addition, the experience from the rotorcraft industry can be brought to bear to help develop guidance material.

- Work on the CBM document currently being developed by the SAE HM-1 committee (JA1013) should be accelerated so that it can engender some useful discussions within the MRO community, both in commercial and the military sectors. This team should involve representatives from the IMRBPB and the MPIG in addition to people from the aviation industry sector. The concept of the P-F interval will form a key element in this as will the specific techniques regarding varying inspection intervals.

Summary/Conclusions

In this SAE EDGE™ Research Report we have considered a few key issues related to IVHM systems and their relationship to maintenance credits. Obtaining a credit for the use of advanced technology is not a straightforward task. Because commercial aviation is such a highly regulated industry, the goal of introducing IVHM systems into aircraft will not be successful without the full involvement of civil certification and airworthiness authorities. Any system that is installed on the aircraft must be certified to be safe and reliable, especially if it is going to drive maintenance processes. As we have outlined in this report, a well-specified process does exist to introduce changes. PHM practitioners should become familiar with this process so they can work with internal and external stakeholders in a coordinated fashion to make this happen. We have shown that the authorities themselves are ready and willing to listen to applicants with such requests. The interest that the IMRBPB and the MPIG organizations have shown in these techniques is very encouraging. With the publication of documents such as AC 29-2C MG15, IP-180, and AC 43-218, the thinking of the FAA is becoming quite clear on the steps that they want applicants to follow. With future modifications of MSG-3, both SHM as well as more general diagnostic and prognostic systems will be considered more favorably than in the past. But the approach needs to be a measured one. The introduction of advanced IVHM systems that will drive maintenance actions will happen in stages. At the beginning, the new and the existing processes will work in parallel. Data from the new IVHM system will be collected and compared with information from the existing system so that the two processes can be compared. This will be used to convince the regulators about the efficacy of the new process. Once the evidence is available and reviewed by the MRB, it will be easier to make the changes permanent.

The application of HUMS in rotorcraft (starting with helicopters operating in the North Sea) gave OEMs a very convenient way to collect critical flight data to enable PHM functions. This allowed them to propose new functionality for getting credits approved. The authorities realized the benefits of this approach and developed guidance to help the industry out. This is a good example to follow in the fixed-wing world as well. Meanwhile, the industry is well on its way to develop novel IVHM techniques to help reduce MMH and overall costs associated with maintenance. For example, the experience with SHM that Delta Air Lines has accumulated will go a long way in helping other applicants of SHM systems. The publication of ARP5987 by the SAE E-32 committee is a milestone in that it explicitly lays out concrete steps that an applicant needs to take to ensure that installing an EHM system or a process is worth doing. Another document that is currently being written in the HM-1 committee will be of interest to the community. This will be a joint aerospace/automobile document known as JA1013 that will examine the linkages between CBM and the MSG-3 process. Much like we have laid out in this report, this document will describe the steps needed to get approval for CBM functions within the commercial aviation space. Such documents will allow SAE International and other standards development organizations to create more detailed guidelines on how to obtain maintenance credits both at the vehicle level and at the subsystem and component levels. The progression will most probably be from individual systems (e.g., brake monitoring, battery monitoring, ODM) to more complicated systems such as LLP parts and then eventually to the vehicle as a whole. This will be a long journey, but every step in this direction will be important; the race will be won, in a slow and steady manner.

SAE EDGE™ Research Reports

SAE EDGE™ Research Reports, like the present report on "Unsettled Issues Concerning Integrated Vehicle Health Management Systems and Maintenance Credits" are intended to push further out into still unsettled areas of technology of interest to the mobility industry. At times, SAE launches these reports before attempting to form a joint working group, let alone a clinical research program or a Standards Committee. At other times, even with existing committees, there are certain topics that need to be highlighted.

SAE EDGE™ reports are intended to be quick, concise overviews of major unsettled areas where vital new technologies are emerging. By definition, an unsettled area is characterized more by confusion and controversy than established order. Early practitioners must confront an absence of agreement. Their challenge is often not to seize the high ground but to find common ground. These scouting reports from the frontiers of investigation are intended merely to begin the process of sorting through critical issues, contributing to a better understanding of key problems, and providing helpful suggestions about possible next steps and avenues of investigation.

SAE EDGE™ Research Reports, therefore, are fundamentally distinct from the more formal working groups approach and far removed from the more mature clinical research program and standard's development process.

Recommendations

The recommendations outlined in this report can be summarized as follows:

- Standards development organizations can take a leading role in generating advocacy for IVHM by publishing guidance material related to specific technologies as well as general principles. Documents such as ARP5987 and ARP6461 (from SAE International) will go a long way to furthering the cause of IVHM in commercial aviation maintenance.

- Intellectual property issues will remain an issue and will have to be tackled fairly. All stakeholders will have to realize that it is in their interest to collaborate and any way of facilitating this in a neutral manner would help. Again, nonpartisan organizations can play a leading role in moving this forward, say by forming consortia to facilitate collaboration with the industry stakeholders.

- The industry will benefit from the publication of AC 43-218 in 2020. This will lay out a set of overall guidelines for IAHM systems. To make these systems more useful, additional guidance for individual technologies will have to be worked on by standards development organizations. Guidance can be sought from existing documents like IP-180.

- It is important for individuals or teams associated with MPIG and IMRBPB to write CIPs or other documents that can influence revisions to MSG-3 guidelines. This is a good systematic way to bring about a favorable change to maintenance guidelines. MSG-3 guidelines are revised every three years, and the next revision is scheduled to be published in 2021. Standards development organizations should be involved in this because much of the expertise for these concepts resides in individuals who are already writing related documents as part of technical committees within these organizations.

- AC 43-218 seeks to give high-level guidance on the certification of health management systems. It is not as detailed as the miscellaneous guidance (AC 29-2C MG15) that is available to the helicopter industry, but it is a very important first step and demonstrates that FAA is very keen on seeing these systems be operational in a Part 25 aircraft. This is a beginning, but more guidance will be needed to make IVHM a viable part of MRO procedures.

- Many lessons can be learned from the rotorcraft industry and engine and airframe OEMs in the commercial aviation should study these lessons.

- The use of COTS equipment (hardware and software) in health management system is inevitable, and the sooner the industry figures out how to deal with their use the better. Short of demanding the source code and testing every line, designers will have to come up with advanced testing methodologies that ensure that the equipment cannot malfunction in any possible scenario (within the parameters of uncertainty that are laid out by the recommended practices acceptable to the industry).

- One of the most expensive parts in an aircraft is an engine LLP, and maximizing its useful life is a worthy goal. Derated operations get us part of the way there, but to get the most benefit, an application of UBL is necessary. Standards development organizations can play a key role in this by writing guidance documents or even information reports regarding this area. As a start, the SAE E-32 committee could begin working on a successor to AIR1872B. The recommendation would be to revive

AIR1872B with a more detailed discussion on LLP lifing and how a true UBL can be put in place.

- The ODM systems described in this section that are already installed on airborne and ground-based applications are producing a lot of data. If this data could be reviewed along with actual condition of the equipment from tear downs, it could provide vital information to determine if these are sufficient to eliminate MCD inspections.

- Scheduled-SHM has been part of the MSG guidance since 2009. The AISC-SHM committee should be working towards including automated-SHM in the next revision of the MSG-3, which will be published in 2021.

- Work on the CBM document currently being developed by the SAE HM-1 committee (JA1013) should be accelerated so that it can engender some useful discussions within the MRO community, both in the commercial and the military sectors.

Abbreviations/Definitions

A4A - Airlines for America

AC - Advisory Circular

AHM - Aircraft Health Management

AI - Artificial Intelligence

AIR - Aerospace Information Report

AISC-SHM - Aircraft Industry Steering Group for Structural Health Monitoring

AMP - Approved Maintenance Plan

ARP - Aerospace Recommended Practice

ATA - Air Transport Association

CBM - Condition-Based Maintenance

CIP - Candidate Issue Paper (IMRBPB)

COTS - Commercial Off-The-Shelf

CVM - Comparative Vacuum Monitoring

DAL - Development Assurance Level

DoD - Department of Defense

EASA - European Aviation Safety Agency

FAA - Federal Aviation Administration

FDAL - Functional Development Assurance Level

FH - Flight Hours

HCF - High-Cycle Fatigue

HUMS - Health and Usage Monitoring System

ICA - Instructions for Continued Airworthiness

IDAL - Item Development Assurance Level

IMRBPB - International Maintenance Review Board Policy Board

IP - Issue Paper (IMRBPB)

IVHM - Integrated Vehicle Health Management

LCF - Low-Cycle Fatigue

LLP - Life-Limited Parts

MCD - Magnetic Chip Detectors

ML - Machine Learning

MMH - Maintenance Man-Hours

MSG - Maintenance Steering Group

MSI - Maintenance Significant Items

MPD - Maintenance Planning Document

MPIG - Maintenance Programs Industry Group

MRB - Maintenance Review Board

MRBR - Maintenance Review Board Report

MRO - Maintenance, Repair, and Overhaul

NAA - National Airworthiness Authority

NDT - NonDestructive Testing

ODM - Oil Debris Monitoring

OEM - Original Equipment Manufacturer

PHM - Prognostics and Health Management

RCM - Reliability-Centered Maintenance

RUL - Remaining Useful Life

SAC - Sikorsky Aircraft Corporation

SHM - Structural Health Management/Monitoring

SSI - Structural Significant Items

STC - Supplemental Type Certificate

TC - Type Certificate/Certification

UBL - Usage-Based Lifing

Acknowledgments

While I am responsible for putting this report together, I received a lot of help from some of the most experienced engineers in the aerospace industry. I acknowledge them alphabetically, starting with Bal Annigeri, who retired recently from Pratt & Whitney. He was an Associate Director responsible for structures and lifing. Mark Davis recently retired from Sikorsky Aircraft, as Tech Fellow for Aircraft

Health Management (AHM), with responsibilities for development and transition of advanced AHM technologies, including methods and processes for using data from existing health and usage monitoring systems to obtain maintenance credits for helicopters. Bill Heliker is with the FAA and is also the current chair of the IMRBPB, which gives him a unique perspective on the issues discussed in this report. Ian Jennions is the Director of the IVHM Center at Cranfield University where he has been leading the effort to develop and disseminate knowledge about IVHM in aerospace and other industries. Marcus Labay from the FAA is a senior certification specialist within the Flight Standards group and is the primary author of AC 43-218, which is a significant initiative that the FAA is making in the application of IVHM for commercial aviation. David Piotrowski is Senior Principal Engineer with Delta TechOps responsible for many of the innovations that they are doing with condition monitoring, including the work on CVM described here. Pierre-Charles Rolland, who serves as Head of Aircraft Operability at Airbus Commercial Aircraft in Toulouse, provided important insights into the relationship between current continued airworthiness instructions and a future where condition-based maintenance could be a viable alternative. Guilherme Torres, who works for Embraer Services & Support within their Maintenance Programs Engineering, has been developing remote monitoring methodologies for their aircraft, getting ready for the day when IVHM becomes a more accepted part of maintenance practices. Rhonda Walthall, from Collins Aerospace, now called Raytheon Technologies, has worked for many years in the commercial aviation sphere at an airline, a manufacturer of auxiliary power units, and is now leading their efforts in the application of digital data in the aerospace industry. She is also the founding chairperson of SAE's new technical committee on Electronic Transactions for Aerospace (G-31).

Without their input and ready accessibility, this report would not have been written or would have been much less informative. Finally, I must thank Monica Nogueira from SAE International, who was the impetus behind this work, and her constant "encouragement" made sure that it was completed in time for publication! William Kucinski, also from SAE, helped out immensely during the editing process, and the report is that much better for his additions and corrections.

Bibliographic References

1. Adams, D.O. and Zao, J., "Searching for the Usage Monitor Reliability Factor Using an Advanced Fatigue Reliability Assessment Model," in *Proceedings of the American Helicopter Society 65th Annual Forum*, Grapevine, TX, May 2009.

2. Beale, R. and Davis, M., "Application of Rotorcraft Structural Usage and Loads Monitoring Methods for Determining Usage Credits," FAA Report DOT/FAA/TC-15/10, 2015.

3. Beale, R.J., Davis, M.W., Kloda, J.R., and Templeton, D.K., "S-92® Main Rotor Blade Life Extension," in *Proc. AHS 72nd Annual Forum*, West Palm Beach, FL, May 2016.

4. Dirgo, R., Rajamani, R., Burkhalter, K., Adams, J.H. et al., "Aircraft Brake and Tire Condition Diagnosis and Prognosis," U.S. Patent Application, 2018/0290639, Oct. 2018.

5. Eaton, "Electronic Oil Debris Monitoring System," Retrieved from eaton.com, Oct. 2019.

6. Federal Aviation Administration, "Certification of Normal Category Rotorcraft," AC 29-2C, July 2018.

7. Federal Aviation Administration, "Operational Authorization of Integrated Aircraft Health Management Systems," Draft Advisory Circular, AC43-218, Mar. 2020.

8. Federal Aviation Administration, "Maintenance Review Boards, Maintenance Type Boards, and OEM/TCH Recommended Maintenance Procedures," AC 121-22C, Aug. 2012.

9. Federal Aviation Administration, "System Design and Analysis," AC 25.1309-1A, June 1988.

10. Foote, P., "New Guidelines for Implementation of Structural Health Monitoring in Aerospace Applications," *SAE Int. J. Aerosp.* 6(2):525-533, 2013, https://doi.org/10.4271/2013-01-2219.

11. Gastops, "ODM," Retrieved from gastops.com, Oct. 2019.

12. International Maintenance Review Board Policy Board (IMRBPB), "Definition of Structural Health Monitoring (SHM)/Addition to MSG-3," Issue Paper (IP) 92, Apr. 2009.

13. International Maintenance Review Board Policy Board (IMRBPB), "Further Advanced Definition of Structural Health Monitoring (SHM)/Addition to MSG-3," Candidate Issue Paper (CIP) 105, Recommended for implementation, Apr. 2010.

14. International Maintenance Review Board Policy Board (IMRBPB), "Aircraft Health Management (AHM) Integration in MSG-3," Issue Paper (IP) 180, Apr. 2018.

15. Jennions, I.K., editor, *Integrated Vehicle Health Management: Perspectives on an Emerging Field* (Warrendale: PA: SAE International, Sept. 2011).

16. Malere, J. and Santos, L., "Challenges for Costs and Benefits Evaluation of IVHM Systems," *SAE Int. J. Aerosp.* 6(2):484-491, 2013, https://doi.org/10.4271/2013-01-2183.

17. National Safety Council, "Injury Facts: Motor Vehicle," https://injuryfacts.nsc.org/motor-vehicle/overview/introduction/, Retrieved Oct. 2019.

18. Nowlan, S.F. and Heap, H.F., *Reliability-Centered Maintenance* (Los Altos, CA: Dolby Access Press, Dec. 1978).

19. Piotrowski, D., Roach, D., Linn, J., Reeves, J. et al., "Validation of a Structural Health Monitoring (SHM) System and Integration into an Airline Maintenance Program (Part 2)," in *A4A 2014 NDT Forum Presentations*, Burlingame, CA, Sept. 2014.

20. Radio Technical Commission for Aeronautics (RTCA), "Software Considerations in Airborne Systems and Equipment Certification," RTCA DO-178C (EUROCAE ED-12C), Dec. 2011.

21. Radio Technical Commission for Aeronautics (RTCA), "Hardware Considerations in Airborne Systems and Equipment Certification," RTCA DO-254 (EUROCAE ED-12C), Dec. 2011.

22. Rajamani, R., *Condition-Based Maintenance in Aviation: The History, the Business and the Technology* (Warrendale, PA: SAE International, Dec. 2018).

23. Rajamani, R., *Diagnostics and Prognostics of Aerospace Engines* (Warrendale, PA: SAE International, Nov. 2018).

24. Rajamani, R. and Waters, N., "Certification of Engine Health Management Systems: Guidelines for Selecting Software Assurance Level," SAE Technical Paper 2011-01-2704, 2011, https://doi.org/10.4271/2011-01-2704.

25. Rolland, P.C., "CBM and MSG-3," Airbus Issue Paper, July 2019.

26. SAE International, "Guide to Life Usage Monitoring and Parts Management for Aircraft Gas Turbine Engines," AIR1872B, Cancelled Sept. 2011, but available from SAE International.

27. SAE International, "Guidelines for Development of Civil Aircraft and Systems," ARP4754A, Dec. 2010.

28. SAE International, "Guidelines and Methods for Conducting the Safety Assessment Process on Civil Airborne Systems and Equipment," ARP4761, Dec. 1996.

29. SAE International, "A Process for Utilizing Aerospace Propulsion Health Management Systems for Maintenance Credit," ARP5987, Dec. 2018.

30. SAE International, "Guidelines for Implementation of Structural Health Monitoring on Fixed Wing Aircraft," ARP6461, Sept. 2013.

31. SAE International, "Determination of Cost Benefits from Implementing an Integrated Vehicle Health Management System," ARP6275, July 2014.

32. SAE International, "Diagnostic and Prognostic Metrics for Engine Health Management Systems," AIR7999 Draft, Dec. 2019.

33. Wade, D., Tucker, B., Davis, M., Knapp, D. et al., "Joint Military and Commercial Rotorcraft Mechanical Diagnostics Gap Analysis," in *AHS 73rd Annual Forum*, Ft. Worth, TX, May 2017.

Contact Information

EDGEresearch@sae.org.

The authors of this document together with the SAE Team responsible for its creation join in expressing our deepest appreciation to all of the individuals who contributed.